P9-APG-644

SCIENCE VERSUS PSEUDOSCIENCE

"I THINK YOU SHOULD BE MORE EXPLICIT HERE IN STEP TWO."

SCIENCE VERSUS PSEUDOSCIENCE

BY NATHAN AASENG

An Impact Book

Franklin Watts
New York/Chicago/London/Toronto/Sydney

Photographs copyright ©: Sidney Harris: p. 2; UPI/Bettmann Newsphotos: pp. 10, 25, 50, 81, 119; Archive Photos, NYC: pp. 23, 41 (both Kean), 118 (London Daily Express); Wide World Photos: pp. 24, 55, 76, 105; The Bettmann Archive: pp. 48, 88, 106; Gamma-Liaison, Inc.: pp. 65 (Shahn Kermani), 90 (Tom Smart), 109 (Alain Morvan); Randy Matusow: p. 70; Reuters/Bettmann Newsphotos: p. 72; Photo Researchers, Inc./Odette Mennesson-Rigaud: p. 78; Hansen Planetarium/Charles Capen: p. 113.

Library of Congress Cataloging-in-Publication Data

Aaseng, Nathan.
Science versus pseudoscience?
by Nathan Aaseng.
p. cm.
(An Impact book)
Includes bibliographical references and index.
ISBN 0-531-11182-2
1. Science—Miscellanea—Juvenile literature. [1. Science—Miscellanea.] I. Title.
Q163.A1018 1994
500—dc20 93-30014 CIP AC

Printed in the United States of America
6 5 4 3 2 1

CONTENTS

ONE

FALSE INFORMATION

Step right up, folks! What you are about to hear is going to change your life forever, and maybe save your life and the lives of your loved ones. You have never seen anything like the astounding science of CHRONOBIOLOGY! *Scientifically proven in test after test, this amazing new discovery can connect you to the natural circadian rhythms that govern every day of your life. By carefully measuring chemical and hormone levels in your body, the magic of* CHRONOBIOLOGY *can tell you when your body is operating at peak efficiency.*

Don't delay another moment! Find out what time of the day is best for making those major money decisions! Learn when you can best memorize your notes for that big test! Yes, CHRONOBIOLOGY *can mean the difference between life and death by telling you when your body is ripe to respond to a medical treatment and when it will not!*

Does this concept of "chronobiology" sound a bit farfetched? How about a slightly different topic of study concerning the mysteries of human body cycles: biorhythms. There is no question that the human body does respond to internal cycles. As evidence, we need only look at our own night-day patterns of sleep and wakefulness, or at menstrual cycles. The study of biorhythms has uncovered remarkably consistent internal cycles that govern a person's ability to perform physically, mentally, and emotionally.

These patterns begin at birth and each runs for a fixed number of days. The physical cycle, for example, has been found to last twenty-three days. By charting a person's biorhythmic cycles, scientists can predict on which days a person will be functioning at his or her lowest level of efficiency. Such knowledge can be especially useful in helping to prevent accidents. Scientific studies have determined that accidents are more than twice as likely to occur during a person's "down" periods than at any other time.

WHERE'S THE SCIENCE?

One of the two "sciences" mentioned above is a legitimate biological research field, operating within the rules of science. The other is an impostor posing as a science. A doctrine, belief, or fraud that is passed off as a science is called a "pseudoscience." "Pseudo" comes from the Greek word for "false." Supporters of pseudoscience wrap their beliefs in scientific clothing to take advantage of the great authority that science carries in our society, even though what they practice has little to do with the scientific process.

Some practitioners are genuinely convinced that their pseudoscience is a legitimate, valuable science; others are merely con artists trying to make a fast buck. Astrology, ESP (extrasensory perception), UFOs (unidentified flying objects), and miracle medical cures all commonly fall within the domain of pseudoscience.

Pseudoscience has been around as long as science itself and its roots go back to superstitions from before the dawn of history. But it has enjoyed increasing popularity in recent years. A Gallup poll found that the number of American youth who believe in astrology jumped from 40 percent to 58 percent between 1978 and 1988.[1] These gullible youth were joined by First Lady Nancy Reagan, who had her husband's daily schedule cleared with an astrologer during Ronald Reagan's presidency. The president did not discount the whole business, admitting, "I don't know enough about it to say if there is something there or not."[2]

Many Americans could identify with Senator Claiborne Pell of Rhode Island, who publicly expressed fears that the Russians would gain a crucial military advantage over the United States because of their superior understanding of paranormal powers.[3]

The inroads made by pseudosciences appear throughout society. According to a 1990 University of Chicago poll, 42 percent of the adults in the United States believe they have been in contact with someone who has died.[4] Many police investigators hire clairvoyants to help them track down clues in important cases. Americans currently spend billions of dollars per year on worthless medical cures peddled by pseudoscientists.

HOW CAN WE TELL SCIENCE FROM PSEUDOSCIENCE?

Obviously, despite living in a world dominated by science and the technology it has spawned, a great many people have trouble telling a science from a pseudocience.

To return to our original example, which of the two fields, chronobiology or biorhythms, is a science, and which is a pseudoscience? Descriptions of the two sound so similar that the average person may not be able to tell real science from the impostor. For, as we shall see, the scientific world is not neatly divided into the noble heroes of science and the evil villains of

President Ronald Reagan and his wife, Nancy. During Reagan's administration, the first lady regularly had her husband's daily schedule cleared by an astrologer.

pseudoscience. Nor is phony science as neatly separated from science by subject matter. Legitimate areas of scientific study, such as medicine, are plagued with their share of pseudoscientists. Other areas of study that are widely regarded as pseudoscience, such as extraterrestrial detection, can be examined in a scientific way.

To further complicate matters, the type of language

used in the descriptions of chronobiology and biorhythms was purposely misleading. *Chronobiology is a science, and biorhythms is a pseudoscience* (see chapter 4). A more scientific style was used to describe the pseudoscientific biorhythms to demonstrate that you cannot tell science from pseudoscience simply by which *sounds* more scientific. It is, after all, the business of pseudoscience to sound scientific.

A shrill, carnival huckster style was used to describe the scientific chronobiology to show how even legitimate science can be confused with pseudoscience when scientists wander away from scientific methods and principles.

We live in an exciting age of scientific research in which startling new discoveries are made almost daily. Far more information is gathered than any individual can absorb. How can the average person sift through the constant flow of information and the banter of media hype to sort out true science from the scientifically packaged arguments of pseudoscience?

The task is not easy, nor is there a simple method of separating what is true about our world from what is false. But if we can begin to understand what science is, how it works, and what its limits are, we will be far less likely to fall under the spell of pseudoscience.

Learning to identify pseudoscience is not a trivial matter of splitting hairs over scientific terminology. It is not a case of overreacting to a harmless, amusing pastime.

On a personal level, pseudoscience provides us with false information. Acting on that information can do more than simply waste our time and energy; it can cost us our money, health, and even our lives. It can cause us to make bad decisions that affect our dealings with others. *One of the main purposes of this book is to help you protect yourself from getting ripped off by pseudoscientists.*

In his book *Galileo's Revenge,* Peter Huber details shocking court verdicts, resulting in millions of dollars' worth of awards, made by juries who were fooled by pseudoscientific experts. Court testimony by witnesses with impressive-sounding credentials has been offered to persuade juries that AIDS can be caused by trace environmental pollutants and that cerebral palsy is primarily caused by incompetent physicians. A soothsayer was even awarded $1 million in damages by a jury that believed her story that she lost her psychic powers after undergoing a CT brain scan.[5]

Pseudoscience also presents problems on a larger level. Pseudoscience poisons the vast pool of knowledge that centuries of experimentation has produced. It replaces the rational study of our world, accessible to all, with a primitive dependence on a few "enlightened experts" who see what no one else can. It substitutes a grab bag of magical tricks for the ordered laws and truths of nature. The false information provided by pseudoscience diverts us from choosing a wise course for the future. *Another goal of this book, then, is to equip you to deal with confusing claims of science and pseudoscience so that you can be involved in helping your community and your country.*

TWO

UNDERSTANDING SCIENCE AND PSEUDOSCIENCE

Science is a process of gaining knowledge about the reality that surrounds us. It is a method of determining how things are structured, how they operate, how they operated in the past, and how different parts of reality relate to each other.

Setting out to study reality is an impossibly huge task, and so science is broken up into more manageable courses of study. Biology, for example, is the study of that part of reality composed of living things. Biology can further be broken down, for example, into such areas as neurology (the study of the nervous systems of living things), and ornithology (the study of birds).

All areas of science collect knowledge in a systematic way. The first step is the collection of information through the five senses of sight, touch, hearing, smell, and taste. From this information, scientists use logic and reasoning to form hypotheses, or educated guesses, to explain what appears to be true about the subject.

Scientists then go about testing the accuracy of those hypotheses through experimentation. Even in sciences such as geology, that deal with past events that cannot be tested, methods of collecting information are evaluated through experimentation.

Experimentation is the final judge in science. Regardless of what the scientist believes or hopes to be true, the hypothesis must pass the test or it cannot be accepted as true. If new facts or experiments disprove the hypothesis, the hypothesis is discarded. If the results fail to either confirm or disprove the hypothesis, further observation and experimentation is needed. Only when the results of the tests confirm the hypothesis can this information be added to the store of scientific knowledge. Scientific facts are constantly tested and revised when further observation or experimentation yields new information.

This pattern—observation, logic, and hypothesis, followed by fact-gathering and experimentation to confirm or disprove, followed by impartial acceptance of the result—must be carried out completely in order to be accepted as true science. You do not have to be a scientist to make observations and use your own logic to construct explanations of reality. But unless these explanations are backed up by facts and well-designed experiments, they cannot be included among scientific knowledge.

Pseudoscience shortcuts the scientific method. Pseudoscientists claim to be following the requirements of science but they actually leave out or mess up one or more of the elements.

THE BOUNDARIES OF SCIENCE

Not everything that we call reality can be tested through experiments. Literature, for example, provides huge funds of knowledge that are not subject to experimentation. That does not mean this knowledge is worthless; it simply means the study of literature is not a science. The existence of a loving God who takes a personal interest in humans is not something that can be confirmed through scientific experiments. That does not mean that belief in God is false; it simply means that such matters lie outside the realm of science.

Pseudoscience is careless about observing the boundaries between what is science and what is not. Pseudoscientists may take a subject that is not a science and present it as if it were. Or they take a subject that is a science and use nonscientific methods to arrive at conclusions.

If true science were always done perfectly, it would be easier to draw the line between pseudoscience and science. But scientists, being human, make mistakes. Scientists do not always correctly interpret the results of their experiments, and this allows false "facts" to enter the scientific pool of knowledge. But the fact that a scientific claim turns out to be wrong does *not* necessarily indicate a pseudoscience.

Wilbur and Orville Wright, for example, based the wing design of their first airplanes on airflow and wind-resistance tables developed by scientists. Only after a frustrating series of failures did the Wrights discover that the tables were so full of errors as to be useless.[1] The fact that the "knowledge" proved to be false was not an indicator that the airflow studies were pseudoscience. The basic procedures of scientific experimentation had been followed. The only problem was that mistakes had been made in either the recording of data or the calculations.

This kind of faulty science is not nearly as dangerous as pseudoscience. Faulty science rarely causes anything more than short-lived frustration and occasionally embarrassment to those who made the errors. Unlike pseudoscience, which weaves a confused web of doubt to cover its tracks, faulty science can be quickly exposed and corrected. The Wright brothers simply ran their own experiments to correct the errors and went on to invent the world's first airplane.

Some of the worst confusion in locating the boundaries that separate science from pseudoscience comes in the new frontiers of science. In these areas of study, experimentation is so incomplete that even conscientious scientists have little understanding of what is fact and what is not. They are limited to educated guesses based on observation and logic and a few bits of experimental data. Unfortunately these limits make it very difficult to disprove the claims of pseudoscientists.

As a result, new frontier science can often be mistaken for pseudoscience, especially when it challenges the most commonly held hypotheses of the scientific establishment. The idea that shooting stars were meteorites—flaming pieces of debris that burned out as they entered our atmosphere—was hooted at for many years. Not until the late eighteenth century was enough evidence gathered to confirm the existence of meteorites.[2]

Similarly, many years ago a German meteorologist named Alfred Wegener developed the novel view that the continents of the Earth were in motion. Wegener proposed that the land masses had been joined in prehistoric times and had been drifting apart ever since.

The notion that the solid land on which we lived was actually floating and moving struck most scientists as ludicrous. Their observations and logic told them it simply was not so. Wegener was unable to produce the kind of evidence needed to support his ideas, and so was ridiculed. The fact that he held firm to his beliefs without the required proof strained his hard-earned reputation as a respected scientist.

Not until the 1960s, long after Wegener's death, did geologists develop core-sampling techniques that found evidence to back up his claim. They discovered that chunks of the earth's upper crust (continents and large islands) rode on top of slabs of the lower crust. Scientific and technological advancement had come to prove an idea that had seemed ridiculous.[3]

Innovators in science often cite such examples and claim that proof for their notions is lacking only because their field of study is still in its infancy. They point out that scientists cling stubbornly to established theories and that many of the most successful scientists have been daring adventurers like physicist Albert Einstein who went out on a limb with bizarre, new ideas.[4] They remain convinced that eventually techniques will be developed to prove them right.

The less information available, the harder it is to tell which explanations are plausible and which are not. Without solid evidence to support them, scientists have a difficult job protecting the public from misguided pseudoscientists, harebrained dreamers, quacks, and rip-off artists. The little-known fringes of science, such as the intriguing effects of attitudes and emotions in combating physical diseases, are currently so sparsely explored that they are fertile ground for pseudoscientists.

Even solid, well-documented areas of scientific study can slip into the domain of pseudoscience. The media, which is always looking for exciting angles on stories, is especially susceptible to the temptation to overstate new findings in science. Any time the results of experimentation are exaggerated, when the conclusions spin off into wild speculation and miraculous claims or jump into matters that are not the business of science, science can degenerate into false science.

The study of economics is on solid ground when it provides useful information about the way a society produces goods and services. But it lapses into pseudoscience when an economist claims a surefire method of predicting what the stock market will do. Special diets, acupuncture, and chiropractic all can provide legitimate health benefits. But when they claim to be able to cure everything from dandruff to hemorrhoids, they stray from medical science into pseudoscience.

THE APPEAL OF PSEUDOSCIENCE

Pseudoscience is a holdover from a human condition that was firmly entrenched long before the scientific method gradually began to take hold in the sixteenth century. In prescientific days, new information was gained strictly from observations and conclusions drawn from those observation. While many wise people were thorough in their observations and cautious with their conclusions, others were less patient in compiling observations and let their active imaginations lead them to jump to irrational conclusions. If these people were persuasive speakers, they could gain a large following for their ideas. The ideas became so firmly entrenched that they were regarded as truth. They could not be disputed even if further observations found discrepancies. Since the ideas were "known" to be true, then any observations to the contrary had to be wrong. Shameful episodes such as witch-burnings and pitiful health notions such as the claim that a king's touch could cure disease were often the result.

While science ushered in a more experimental method of exploring the world, it did not quench those active imaginations or cure people of the fault of jumping to conclusions. In fact, according to one historian, "science provided an immensely powerful engine for generating irrational beliefs at will."[5]

This happened because the stunning achievements of science and technology in the nineteenth century gave it an air of superiority. Science was considered the playground of the enlightened, even by people who knew next to nothing about science. All they knew was that the miracles produced by science proved that science had the answers.

In such an environment, any theory that wrapped

itself in the mantle of science could gain instant respectability from a large number of people. Since most people did not know how science worked, pseudoscience could be as persuasive as science. As long as it was couched in scientific terms, people would buy virtually any snake-oil cure that came along. Dozens of scientific "breakthroughs" would sweep over the country in a blaze of popularity, then fizzle as results proved less than expected. As early as 1838, Harriet Martineau complained of the "fickleness of Americans" in rushing from "one scientific fad to another."[6] These included such "sciences" as phrenology (the reading of a person's character based on the contour of the skull) and mesmerism (bringing health by using magnets to manipulate a person's "universal fluid").

Science has retained this lofty status in our time. Scientists have achieved such mind-boggling feats as unleashing the power of the atom, landing an astronaut on the moon, and developing machines that can compute millions of equations in a second. They speak and write in highly specialized jargon that the average person cannot begin to understand. Anyone who doubts the competence of science need only look at skyscrapers and video recorders and thousands of other bits of technology that have altered our world.

Businesses take advantage of this aura of competence by wheeling out bespectacled, white-coated scientists in commercials to sell new products. Lawyers call scientists as "expert" defense witnesses to persuade juries to their way of thinking. Pseudoscientists use the intimidating status of science to persuade the public to accept their notions.

The public today certainly knows more about science than it did a century ago. Yet the pace of scientific research far outstrips the ability of the average person to keep up with it. As a result the general public is hampered by the same ignorance of what actually is going on in science that plagued the scientific fad-chasers of the nineteenth century. According to one American scientific publication, "at least nine of ten citizens lack the scientific literacy to understand and participate in the formulation of public policy on a very important segment of the national agenda."[7] A recent survey conducted by the National Science Foundation was even more harsh. It concluded that just over 5 percent of adults in the United States qualify as scientifically literate.[8]

"AWED BY COINCIDENCE"

A population that reveres science but has little understanding of it is an ideal breeding ground for pseudoscience. A prime example of the way pseudoscience feeds on that ignorance is the phenomenon of coincidence. The average person, according to one scientist, is "awed by coincidence."[9] When something unusual happens for which there is no apparent logical explanation, we become uncomfortable. In the absence of logical explanations, we become open to illogical ones.

For example, how likely is it that you will flip a coin in the air ten times and have it come up heads each time? It would seem to take a pretty bizarre streak of coincidence for that to happen. What if you saw it done by a person who claimed to have the psychic ability to mentally manipulate the coins? Normally you might be highly skeptical of the person's claim to have psychic abilities. But how else do you explain what you have just seen? In light of your observations, you might well conclude that this person has demonstrated telekinetic powers.

A little understanding of odds, chance, and coincidence, however, would tell you that if a person flipped a coin 100,000 times, the odds are that there will be nearly 100 cases in which heads will come up ten times in a row. Far from being a miracle, such a streak is normal and predictable.[10]

In another example, suppose you begin chatting to a complete stranger sitting next to you on a bus. During the course of your conversation you discover that this person's boss is a good friend of your brother. On your next bus trip, you find that your seatmate, again a stranger, was in the army with your next-door neighbor. On a third trip you find that the new pastor at your church once married the parents of the stranger sitting next to you.

Now you are thinking all these coincidences are just too strange. What is going on here? No rational explanation seems available, so you begin wondering if some outside force is at work. A greater understanding of odds, chance, and coincidence, however, would tell you that what happened is not something out of the Twilight Zone but is, in fact, quite expected. The odds are more than 98 out of 100 that any two Americans are linked to each other in some way by two or fewer intermediaries.[11]

To test your understanding of coincidence, how many people must be gathered together in order to *guarantee* that two of those people share the same birth date? Since there are 366 possible birth dates, the answer is 367. How many people must be gathered together to provide a fifty-fifty chance that any two share the same birth date? Since 50 percent is half of 100 percent, it may seem logical that half of 366, or 183, are needed. Actually, if you work out the mathematical probabilities, the answer is 23.[12]

Pseudoscientists take advantage of people's misconceptions of coincidence. Those odd experiences that happen to everyone seem to cry out for explanation. In the absence of a logical, verifiable answer, a totally unsupported explanation, even an illogical one, is often more comforting than no answer at all. In reality, a certain level of spectacular, unusual coincidence is predictable based on the law of averages.

GREAT SCIENTISTS' FLIRTATIONS WITH PSEUDOSCIENCE

Scientific ignorance, though, does not explain why a number of prominent scientists throughout history have dabbled in pseudoscience on the side. Michael Faraday and Lord Kelvin were two respected physicists who believed in the paranormal. Isaac Newton actually wrote more about alchemy—the craft of turning ordinary metals into gold—than he did about physics.[13]

Pseudoscience feeds not only on ignorance but on the fact that the senses are not totally reliable. Most people today are wise to the magic of illusions and the special effects of motion pictures. Hardly anyone today would be as gullible as nineteenth-century author Sir Arthur Conan Doyle. The creator of the master detective Sherlock Holmes was such a poor detective himself that he fell victim to one of the most ridiculous pseudoscientific hoaxes of all time.

Doyle believed in the existence of fairies and seized upon a photograph as evidence. The photograph he cited showed two girls standing behind several tiny, winged humanlike figures that were fluttering above a hedge. The photograph was an obvious hoax, and later the girls who took it admitted to cutting out pictures from a book and dangling them from string.[14]

While we may be more sophisticated then Doyle, even the most intelligent among us are still susceptible to illusions, as any able magician can prove. It is no accident that some of the most vocal critics of pseudoscience have been magicians, such as Harry Houdini and James Randi, who know how easy it is to play tricks on people's senses.

Isaac Newton (1642–1727) is generally acknowledged to be one of the greatest scientists who ever lived, yet much of his work crossed over into the realm of pseudoscience.

PLAYING ON THE PUBLIC'S SUSPICIONS OF SCIENCE AND TECHNOLOGY

Pseudoscience has widened its appeal in recent years by playing both sides of the fence regarding the public's respect for science. While taking advantage of scientific prestige, it also appeals to those who are disillusioned with some of the side effects of science and technology.

Some people, seeing the environmental destruction

Sir Arthur Conan Doyle (1859–1930), creator of the great fictional detective Sherlock Holmes, was a proponent of spiritualism. He is shown here with his collection of "psychic pictures."

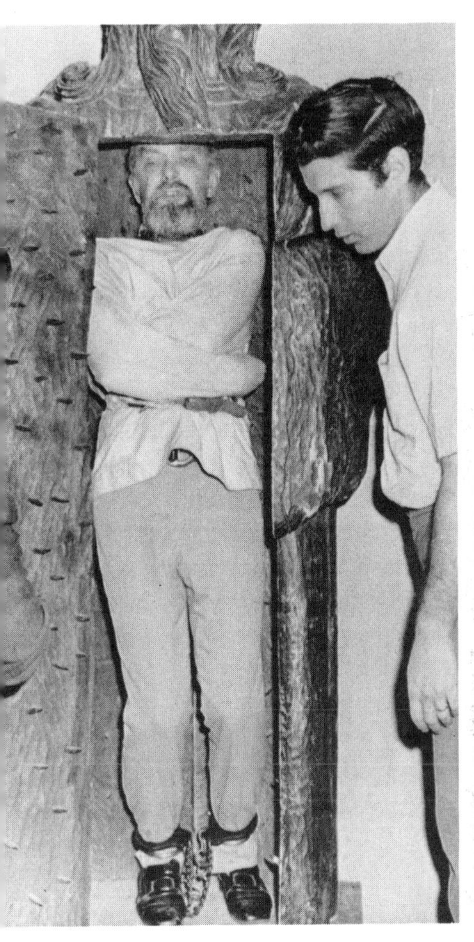

The magician James Randi ("The Amazing Randi") is shown here being placed in a spiked tomb after being straight-jacketed and leg-ironed. The tomb was then tied up with 60 feet of heavy rope. It took Randi only nine minutes to free himself. Randi has also been an outspoken critic of claims made for ESP and other paranormal phenomena.

and the potential for nuclear nightmare caused by science, question whether the miracles of science are a blessing or curse. One Russian observer blames the nuclear disaster at Chernobyl for helping fuel a dramatic rise in pseudosciences in the former Soviet Union.[15] Others lament the emphasis of society on science and technology education at the expense of the arts and humanities. They view science researchers as narrow technicians who ignore the human soul and spirit. Still others cite scientific errors as proof that the scientific establishment does not have all the answers.

Scientists have sometimes fueled this discontentment with a tendency to treat people as statistics on a chart or dots on a graph instead of as human beings. Medical patients, in particular, become irritated at such treatment, rebel against it, and seek help outside the scientific community from more user-friendly practitioners (see chapter 4).

Some arrogant scientists have overstepped their bounds in claiming that science is all that matters in evaluating human existence. They communicate the sense that "today what is real for the most part is what can be proven in the laboratory."[16] This attitude fails to recognize the overwhelming conviction among humans that there is a spiritual side to life that cannot be denied. Recognizing that scientific knowledge does not fill voids in their lives, disillusioned people reject the rational approach altogether. They drift into the more mystical aspects of pseudoscience, such as extrasensory perception, astrology, and crystal healing.

Scientists, being fallible humans, can also be stodgy and can turn a blind eye to new evidence that challenges accepted theories. For years, the scientific community assured the athletic world that anabolic steroids did not have a significant effect on athletic performance. Athletes who used steroids, however, found that the drugs helped them to become stronger and faster. For years the medical community snickered at claims that vitamins and mental attitude fight disease. Now they are finding out that there may be something to those claims after all. These mistaken stances create doubts that open the door for pseudoscience. They fuel suspicion that when scientists scoff at pseudoscience they are simply showing their usual close-minded reaction to new ideas.

Finally, pseudoscience prospers because it is so often based on strong beliefs. These beliefs, which may spring from scientific or unscientific sources, help a person make sense of an often confusing world. They give explanations that provide purpose and establish order and control in the world.

Therefore it is not surprising that many pseudosciences have to do with predicting the future, with providing control over that vast unknown destiny that lies ahead of us. Humans are frequently on the lookout for ways to erase that unknown factor. Sociologists believe that this urge to predict the future and establish controls in the universe becomes especially strong during social unrest, when normal societal traditions are upset.[17]

Strong beliefs are far more difficult to disprove than hypotheses. If you are convinced that your destiny is ruled by the stars, the best scientific logic may not change your mind. If your entire value system depends upon the Earth being no more than 4,000 years old, the best geologic evidence to the contrary may not persuade you otherwise. If you "know" that you have been abducted by aliens, there is virtually no conceivable evidence that could persuade you otherwise.

While nonscientific belief systems are essential in providing an individual with guidance in those matters for which science can provide no answer, their resistance to evidence is incompatible with science. This has become a matter of law in the United States.

When beliefs come into conflict with established scientific fact, such as when seriously ill children are not treated because the parents' beliefs prohibit it, the parents can be held criminally liable.

EXPOSING PSEUDOSCIENCE: TEN CHARACTERISTICS OF SCIENCE

Considering the many layers of camouflage that cover pseudoscience as it hides among the sciences, the task of recognizing pseudoscience may seem hopeless. If we do not have the time or talent to become experts in all scientific fields, if so many scientific fields are left wide open to speculation because of lack of data, if scientific information is not always right, and if we cannot trust our own senses and perhaps not even our own beliefs, how are we to find a dividing line between science and pseudoscience?

There may not be a clear line or an easy test to separate the two, but the task of exposing pseudoscience is not impossible. Even the most scientific-sounding fake gives out telltale clues that it is not the real thing. The following are ten of the most important characteristics of science.[18] Eight of them describe what science is or does; two describe what science is not or does not do.

Any area of study that meets all of these criteria can safely be considered a science. Any area of study that fails to meet one or more of these criteria must be examined more closely to see whether you are dealing with a new frontier science for which information is not completely available, or pseudoscience. The more scientific criteria that are violated, the more obvious it becomes that you are dealing with pseudoscience. Few pseudosciences fail *all* of these scientific criteria, but most pseudosciences will violate several.

TEN IMPORTANT CRITERIA FOR DETERMINING WHAT IS SCIENCE

1. SCIENCE IS LOGICAL AND RATIONAL.
2. SCIENCE MAKES WELL-DEFINED CLAIMS.
3. SCIENTIFIC HYPOTHESES ARE FALSIFIABLE.
4. SCIENTIFIC EXPERIMENTS ARE REPEATABLE.
5. SCIENCE REQUIRES THAT CLAIMS BE EXAMINED BY PEERS.
6. SCIENCE VIEWS UNEXPLAINED GAPS IN THEORIES WITH SUSPICION.
7. SCIENCE REQUIRES CAUTION IN EXAMINING EVIDENCE.
8. SCIENCE REQUIRES OBJECTIVITY.
9. SCIENCE DOES NOT ACCEPT COINCIDENCE AS PROOF.
10. SCIENCE DOES NOT ACCEPT ANECDOTAL EVIDENCE AS PROOF.

CRITERION #1: Science is logical and rational. For all of the bewildering jargon used by scientists, science is primarily based on logic. Its conclusions follow from a logical, rational line of thought. For example, few people would argue with the statement that humans see through their eyes. That is a logical conclusion drawn from, among other things, the observation that when we shut our eyes we can no longer see. A proposal that ears are actually the organs of human sight is not logical. There is no rational line of thought that leads from our observations to that conclusion.

What provides power for our automobiles, burning gasoline or invisible animals pushing on the fender? If you examine an engine and study how it works, you would logically conclude that burning gasoline provides the energy. There is no logical, rational line of thought that leads to the conclusion that invisible animals provide the power.

Pseudoscience is not always restricted to logical, rational thought. The further it strays, the more likely it is to be exposed as an impostor. Take, for example, the argument that aliens in spaceships are abducting people from inner city apartments without being seen by witnesses. Our understanding of the logistics involved in traveling through space and invading a highly populated area without leaving a trace makes this an illogical conclusion. The argument that scientists only think that the Earth is millions of years old because an evil spirit put old rocks on the Earth to fool them and destroy their faith can be accepted only by taking a giant leap away from a logical examination of the evidence.

Claims that wander just slightly from logic are more difficult to judge. There is no known mechanism or principle by which a wooden stick is attracted to underground water, nor is there a logical reason why wood should be attracted to below-surface water when it is not attracted to above-surface water. Therefore the claim that people can detect underground water with divining rods is not a logical conclusion from the evidence. But it is close enough to the concept of magnetic attraction to plant a seed of doubt in many.

CRITERION #2: Science makes well-defined claims.

Good scientific researchers are specific in describing their fields and in making predictions. They use precise terms so that there can be no mistake about what they are saying. Astronomers calculate exactly when they expect a comet to appear in the sky. Chemists calculate exactly what amount of a substance will provide a certain reaction. If something does not go according to plan, the error will be obvious to everyone. Good scientists are willing to go take that risk.

The tendency to be vague in making claims and giving explanations is an indicator of pseudoscience. Be especially careful of those whose predictions of the future use generalized language or mysterious images that can be interpreted a number of ways. Be suspicious of those who will not put themselves in a position where they can be proven wrong.

CRITERION #3: Scientific hypotheses are falsifiable.

This means that the educated guesses made by scientists can be tested to see whether or not they are true. Any hypothesis that cannot be tested is only conjecture and cannot be included among scientific knowledge.

Even research into the past, such as theories about the origins of species, which cannot be examined directly, must be able to collect strong evidence and demonstrate the principles of the theory. Again, this involves a risk among scientists. Their ideas can be proven wrong.

One of the most common characteristics of a pseudoscience is a reliance on hypotheses that cannot be proven wrong. The claim that a certain phenomenon is not detectable by any scientific instrument but only by those with a special gift is an example of an nonfalsifiable hypothesis. What possible evidence could disprove it? Is there any possible evidence that would refute the claim that UFO body snatchers go undetected because they erase the memories of all observers?

CRITERION #4: Scientific experiments are repeatable.

Proper scientific experiments are designed so they can be duplicated by others. They should yield consistent results regardless of who performs the experiment.

Pseudoscientists often support their arguments by citing the astounding results of isolated experiments. But when other scientists attempt to duplicate these experiments, they either cannot get enough information to allow them to conduct the experiments or else do not get the same results.

CRITERION #5: Science requires that claims be examined by peers.

Scientific journals may seem dull and unreadable to the nonscientist. But scientists' reports are written in precise and intricate detail so that other scientists can see exactly what has been done. A good scientist has nothing to hide. He or she wants other scientists to repeat the experiments and weigh the evidence to satisfy themselves that the claims made are accurate.

Pseudoscientists may be vague and secretive in describing their methods. Psychics, especially, are fond of rejecting peer review by claiming that one must have a particular gift to be able to detect the truth of their claims.

CRITERION #6: Science views unexplained gaps in theories with suspicion.

The German bacteriologist Robert Koch set the standard for this criterion in laying down the laws of disease research in the late nineteenth century. Koch insisted that four steps must be followed in establishing whether an organism was responsible for causing disease:

1. That organism must be connected with every case of a particular disease.
2. A pure culture of the organism must be grown outside the body.
3. The disease must be produced whenever this pure culture of the organism is introduced into a healthy animal.
4. As an extra precaution, the organism must be recovered from the diseased animal and cultured again as a check.[19]

While scientists might gather evidence as to the cause of disease without following Koch's rule, only by using the Koch system could they establish an airtight case.

Of course, those exploring new frontiers of science cannot begin to explain everything that is going on in that field. Yet a scientist will make clear the difference between thoroughly researched scientific claims and speculation about the unknown.

Pseudoscientists may make strong claims on the basis of incomplete results. This includes jumping from a small bit of fact to a sweeping conclusion without bothering to account for the unexplained gaps in information. Examples are many health food fads, such as those that the claim that taking certain vitamins will increase sexual potency.

CRITERION #7: Science requires caution in examining the evidence.
Because they accept that any claim they make can be proven false, good scientists are careful about rushing to judgment on promising leads. Bold, new theories are not accepted overnight. Proper scientific debate resists the influence of the media, which are under pressure to produce sensational stories to attract a larger audience.

Most scientific knowledge becomes established fact after painstaking research and verification, which may take several decades. Science operates under the principle that the established knowledge reigns until proven otherwise. *Extraordinary claims must be supported by extraordinary evidence before they can be taken seriously.* Scientists are not required to disprove claims that lie well outside the accepted norms of science. The burden of proof rests with those making the extraordinary claims.[20]

Pseudoscientists often make miraculous claims based on early or incomplete evidence. Rather than exercise restraint, they tend to go public with their exaggerated claims. When challenged they insist that their claims are valid simply because no one can disprove them.

CRITERION #8: Science requires objectivity.
Opinion and established "facts" must be corrected when faced with solid evidence to the contrary.

The famous cases, such as that of continental drift, in which the scientific world discounted ideas that turned out to be correct, are often cited to show that scientists are prejudiced against new ideas. Actually, these cases demonstrate the opposite. While scientists are skeptical of radical new ideas, these instances showed that eventually the mounting evidence caused scientists to reevaluate their previous conclusions. In science, those ideas that are supported by fact eventually win out over even the most accepted traditions.

Pseudoscientists are often champions for their cause. They may have a personal stake in the outcome of research, or their beliefs may be so strong that no conceivable evidence would make them change their minds. When the facts start to go against them, phony scientists may change the procedure or the data rather than their minds, or select only the data that support their position.

CRITERION #9: Science does not accept coincidence as proof.

Suppose a harmless bacterium happened to be found in many animals with rabies. From this rather strong coincidence one might easily conclude that the bacterium causes rabies. Koch eliminated that mistake by requiring that a pure culture of that bacteria be grown and injected into a healthy animal to see if it actually produces the disease.

Scientists understand the nature of coincidence and take steps to minimize its effects. Pseudoscientists may offer alternative explanations for coincidence, may fail to take the laws of probability into account, or may even accept coincidence as proof.

CRITERION #10: Science does not accept anecdotal evidence as proof.

Many pseudoscientific claims rely almost completely on stories and testimonials as proof. People recount incidents that happened to them to prove that a certain method of healing works or that out-of-body or psychic phenomena exist.

A number of sciences, particularly psychology, make use of anecdotal evidence to gather information. Anecdotal stories, after all, report information gathered by the senses. But scientists treat anecdotal evidence with caution. These stories are isolated reports that usually cannot be confirmed or reviewed. The details are sketchy and leave many variables unaccounted for. Also, as any magician knows, people's senses can be misled into making them think something happened that did not.

THE HUNTERS

The criteria discussed above offer a tool for measuring whether or not a scientific claim is truly following the rules of science. A very basic example highlighting some of the variations of science and pseudoscience discussed earlier shows how the clues can be put together to evaluate a claim.

A group of villagers comes across a set of animal tracks near their home. Hunter A determines that the track was made by a deer traveling east. Hunter B determines from experience that the tracks were made by a wolf traveling west.

Hunter A produces a deer foot so that his fellow hunters can see the sort of print produced by a deer and can determine for themselves which end of the print was made by the forward part of the hoof. Hunter B does the same with a wolf foot. Upon examination everyone, including Hunter B, can see that Hunter A was correct.

Hunter A practiced legitimate science. This hunter's conclusion was supported by strong evidence. None of the ten science criteria was violated.

Hunter B also practiced legitimate science even though his conclusion was found to be unsupported by evidence and was clearly demonstrated to be incorrect. The hunter is not guilty of pseudoscience because the basic rules of science were followed. Hunter B simply made a mistake that was easily corrected.

Hunter C determines that the track is twelve hours old, based on her study of how the edges of prints deteriorate with time. Hunter D determines that the track is twenty-four hours old, based on her observations of how the compression of the sand varies with time. Neither provides enough evidence to convince the majority of hunters.

This situation presents a couple of suspicious characteristics. The evidence is not thorough (criterion #6). Since they cannot both be right, at least one of the hunters has not shown proper caution in examining the evidence (criterion #7). *This suggests new-frontier science.* As with Hunter A, rational guesses have been made based on observation. But unlike the first example, there is not yet enough evidence to support one or the other. New-frontier science is the most difficult to distinguish from pseudoscience.

Hunter E, aware that the village is starving, declares that the deer was sent by God in answer to prayer. A glaring violation of scientific criteria shows up here—the nonfalsifiable hypothesis (criterion #3). No experiment or evidence can prove the statement false. That is a strong indication that this is not a science but *a conclusion based on a nonscientific belief system.* Science can make no judgment whether the statement is true. Any use of scientific argument claiming to prove or disprove the conclusion is pseudoscience.

Hunter F declares that the distance between tracks indicates that hunting will be good for the next month because the last two times he saw deer tracks so spread out, the village had no major food shortage. This situation fails to meet a number of scientific criteria. Although the hunter may be convinced he is right, he has shown no logical connection between evidence and conclusion (criterion #1). What logical connection exists between the distance between tracks and the abundance of food? His claim is vague (criterion #2). What exactly does he mean by "good" hunting or no "major" food shortage. The evidence consists solely of coincidence (criterion #9) and personal anecdote (criterion #10). This is pseudoscience.

Hunter G says that the irregular pattern of the tracks shows that the deer will bring bad luck to anyone who eats it. Again, this falls short of many scientific criteria. No logical connection is provided linking irregular tracks to bad luck (criterion #1). No evidence is provided, nor any clue as to how the conclusion was reached, violating scientific criteria 4, 5, 6, and 7. If the hunter's motive is to persuade the other hunters to leave the deer alone so that he can have it for himself, this is fraud as well as pseudoscience. Fraud occurs when people manufacture "scientific" information for personal profit.

Real life, of course, provides far more complex battlegrounds on which scientists and pseudoscientists struggle to capture human minds. It is beyond the scope of this book to present all the evidence, pro and con, for each science and pseudoscience. The expanding nature of science and the faddish nature of pseudoscience guarantee that the evidence is constantly changing.

Rather, the following chapters will show how use of the various scientific criteria will provide some tools for disarming pseudoscience, no matter what shape it may take.

THREE

BATTLEGROUND: MIND AND SPIRIT

In 1898 a fiction book was published about the voyage of a luxury ocean liner called the *Titan*.[1] The tragic story told how the *Titan* struck an iceberg and sank in the Atlantic Ocean, with great loss of life. Fourteen years later, a luxury ocean liner called the *Titanic* set out across the Atlantic Ocean, struck an iceberg, and sank with great loss of life. An eerie coincidence, or did that fiction author have a glimpse into the future?

A man senses a bank robbery taking place and alerts the police, who arrive in time to nab the outlaws. A doctor suddenly feels needed by a patient and visits that person just in time to save his life. A woman about to board an airplane gets an uncomfortable feeling and decides to cancel her flight. The airplane then crashes on takeoff. A young couple is terrorized by some mysterious evil presence in a house where a brutal murder once occurred. The telephone rings and you know who is calling before you pick up the receiver. A dream about one of your relatives comes true.

How does science explain these occurrences? Is there a whole realm of life that exists outside the clearly definable physical world? Do we have some kind of extra sense that works outside the normal limits of behavior to tap into that other world?

Matters involving the mind and spirit present by far the most difficult problems in separating science from pseudoscience. The intelligent mind is bewilderingly complex. The vast range of human choice, and the almost infinite variety of experiences upon which those choices are based, make properly controlled experiments in psychology difficult to design. This complexity means that psychology, the study of the mind, operates under some restrictions. While psychology can teach useful things about how the mind works, it is not as exact a science as chemistry or physics. One atom of a chemical reacts to a condition in the same way as another atom of that chemical. A chemist can predict the result with accuracy. Two humans subjected to identical conditions, however, may react in different ways. A psychologist cannot predict the result with the same kind of accuracy.

The complexity of their subject requires those who study the human mind to occasionally fill in the many gaps in their data in ways that do not fit the definition of science. Sigmund Freud, one of the most influential figures in the early history of psychology, roamed well outside the established rules of science. His use of dream interpretation in psychoanalysis did not set out clearly defined relationships between data and conclusion (criterion #2), but rather relied on an individual analyst's skill in sorting out vague connections. Many of Freud's notions were nonfalsifiable hypotheses (criterion #3). How could anyone disprove his division of the personality into three parts: the ego, the superego, and the id? His explanation that overly strict toilet training could produce either messy, wasteful habits in a child triggered by resentment, or meticulous, orderly habits triggered by compliance, can explain any data, no matter how contradictory.[2]

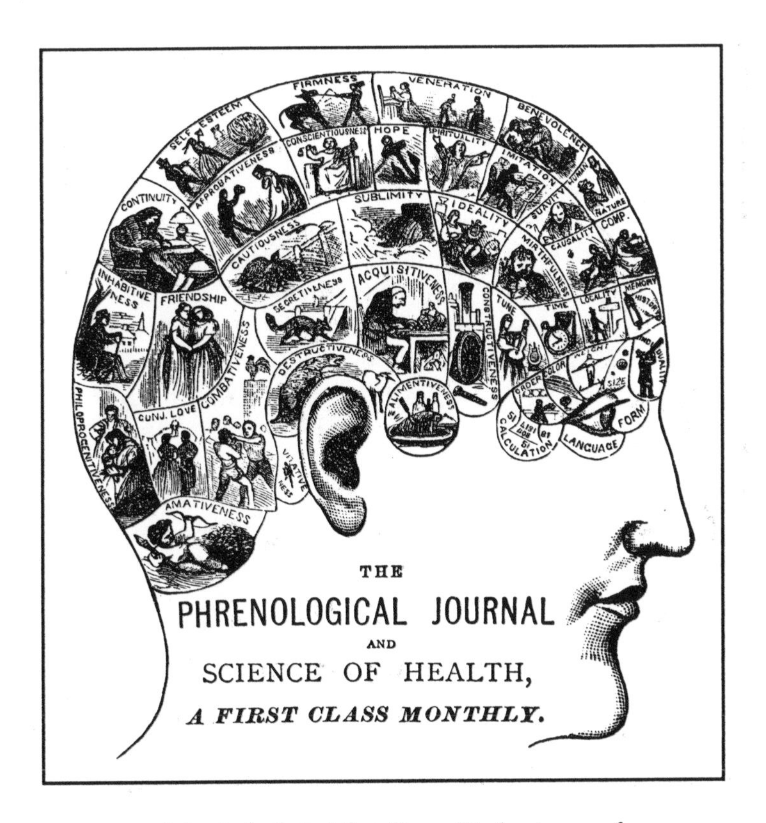

A basic belief of the discredited science of phrenology was that the shape of the skull served as an indicator of mental faculties and character.

DO OUR SENSES DECEIVE US?

Another problem with studying the mind is that the mind must base its responses on the data it receives from the senses, which can occasionally be unreliable. Optical illusions and false recognition cause our eyes to report information that is not true. Memories fail, causing us to believe information that is not true. Occasionally we hear noises for no apparent reason.

Further, humans make errors in drawing conclusions from the data accumulated by the senses. Logic can be faulty. Rational thought can be overridden by emotions or by respect for someone whose arguments cause one's own conclusions to be abandoned. All of this makes it difficult to predict what a person will do, or to understand why the person is doing it.

If the study of the mind presents difficulties for science, the study of the spiritual world is even more tricky. Most religions teach that the universe is ordered by a divine being (or beings) beyond our ability to understand, whose power is greater than the natural laws that control our world.

What does science have to say about this? Nothing. God is a concept that is not confined to the limitations of the physical world. Anything that is by definition beyond our means to detect or measure lies outside the realm of science and cannot be evaluated according to scientific criteria. God is a concept that can only be understood in ways that do not involve science. There is no scientific way of gaining data on which to base a scientific evaluation.

The effectiveness of prayer in influencing the future is a similar subject that cannot be scientifically measured. Proper controls cannot be provided. Prayer cannot be disproved because if something you pray for does not come true, that may be God's negative answer to the prayer. It cannot be proved because there is no way to demonstrate that the event would not have happened without the prayer.

ESP and the occult are considered paranormal phenomena—things that exist outside our normal, scientific understanding of the world. They are similar to religion in that they offer explanations for things that lie outside the direct evidence of the physical world. In fact, some sociologists describe paranormal beliefs as alternatives to religious beliefs, and trace the boom in paranormal beliefs in the 1960s to a decline among mainstream churches.[3] If the realm of religion lies outside the bounds of science, then obviously a great deal of the paranormal does, too.

Yet attempts have been made to measure paranormal activity by scientific means, and these are fair game for applying the scientific criteria. So, too, is any logical evidence produced in support of any paranormal phenomenon.

ESP

Because it overlaps areas that have little to do with science, the study of ESP can provide fertile ground for pseudoscience.

A majority of Americans, as many as two-thirds according to one poll, believe in the existence of ESP.[4] More than half claim to have had at least one experience that they attribute to ESP. Supporters include many of the best educated, including one third of academic professors in one poll,[5] and such noted scientists as Lord Kelvin, Thomas Huxley, and Michael Faraday. Psychics were used by the American government in the 1980s to study Panamanian dictator Manuel Noriega and to seek out the terrorists responsible for killing 241 American soldiers in Beirut in 1983. Military advisers in 1986 urged the training of psychics for duty. Respected American universities include parapsychology (the study of ESP and similar phenomena) in their course offerings.

ESP is the ability to gain knowledge in a way that does not involve the normal senses. It can be divided into several more specific subjects. *Precognition* is the foreknowledge of events that have not yet taken place. *Clairvoyance* is the ability to see objects that cannot be seen through normal vision. *Telepathy* is the ability to communicate without spoken or visual messages. *Psy-*

chokinesis is the ability to move or manipulate objects through mental powers.[6]

The first real scientific effort to study ESP was performed by J.B. Rhine at Duke University in North Carolina, beginning in the 1930s. In his most famous experiment, Rhine recorded the ability of various people to guess which of five symbols was drawn on a hidden card. Rhine's experiment used a total of twenty-five cards, five cards for each symbol.

During the 1930s Rhine provided evidence that some individuals had a special ability, which he called extrasensory perception (ESP), to know what was on the cards without actually looking at them. His most successful subject once correctly identified 119 of 300 cards shown to him. The statistical odds of that happening were so ridiculous that Rhine concluded it could not have happened by chance. Even modern skeptics have had to admit that "subjects trying to guess targets have obtained scores that cannot be attributed to chance."[7]

Does this scientifically demonstrate the existence of ESP? To find out, we need to measure the evidence against our scientific criteria. Scientific support of ESP immediately runs into a host of problems with the question of logic (criterion #1). There is nothing logical about ESP. In more than fifty years of research, scientists have failed to find any receptors or senders in the brain for transmitting ESP, and have been unable to develop any instrument that can measure its presence.

That could simply mean that researchers are having trouble breaking into a new frontier science. But Rhine's experiments run into another barrier. While Rhine correctly followed the procedures of science in setting up a simple test that could easily be repeated by others, only a few researchers have been able to duplicate his findings. Most trials similar to the ones performed by Rhine have yielded nothing but chance scores.[8] The gambling industry is one giant, long-running experiment testing the ability of people to know what is on cards without reading them. Casinos, gambling halls, and lotteries risk millions of dollars each day; in essence, they are betting that people will not be able to know what numbers are on certain cards without seeing them. If ESP were a factor, then certain individuals would be continually winning fortunes. Gambling institutions would lose money hand over fist and would close down. Instead, gambling follows the law of averages and remains profitable.[9]

This lack of repeatability of Rhine's experiments violates criterion #4. That brings into question the controls used in the original. Could Rhine's subjects see through the cards? Was something tipping them off? But even if his controls were proper, the lack of repeatability means he did not scientifically demonstrate the existence of ESP.

Some ESP supporters explain this by saying that ESP is an unconscious, spontaneous process.[10] It comes and goes randomly and cannot be produced on demand. But this is a nonfalsifiable hypothesis (criterion #3) that can explain away any possible experimental result.

Then how does science explain Rhine's puzzling results? It cannot offer ironclad proof that ESP does not exist. There is no possible experiment that could prove that ESP is not real. But it is one thing to say that no scientific explanation has been found, and quite another to say that there is scientific proof of ESP. The lack of repeatability in Rhine's experiments are ample evidence that they do *not* prove ESP.

Experiments known as the "ganzfeld" experiments have often been used as evidence that ESP exists. These experiments isolate a subject in a dark, quiet place free from any visual stimulation. While isolated the subject is tested to see how well he or she can receive extrasensory information. For example the subject is asked to "receive" an image from a randomly selected photograph. The subject then describes thoughts, feelings, and images that are going on in his or her mind. A number of studies have reported that a significant number of people have been able to verbalize information gained from the pictures they cannot see.

Roy Hyman of the University of Oregon examined all the available reports of ganzfeld experiments, forty-two studies in all. He found flaws in all the experiments. Defects included poor documentation, errors in statistical methods, and inadequate safeguards. Hyman found that the actual probability of a positive response in experiments was five times greater than reported. He concluded that "ganzfeld, despite initial impressions, is inadequate either to support the contention of a repeatable study or to demonstrate the reality" of ESP.[11]

The United States Army has a keen interest in determining the validity of ESP. Extrasensory powers could have an enormous impact on military matters. They could be used to gather intelligence, to telepathically destroy weapons and jam computers. The Army Research Institute asked the National Academy of Sciences to form a committee to explore the possibilities of using ESP to improve American military effectiveness.[12]

The Academy examined a number of claims made about ESP. For instance they looked at experiments cited by parapsychologists that indicate a slight ability of people to influence the selection of random numbers. In millions of trials run by researchers Robert Jahn and Helmut Schmidt, subjects tried to telepathically influence a computer-generated sequence of random numbers to change into a nonrandom sequence. Their results have shown between .05 percent and .004 percent more success than would be expected by chance. An evaluation of these experiments by the Academy found that even this meager success was questionable. The report concluded that only one experiment in fifteen years of research showed even a "marginally significant" result in support of ESP.

The National Academy of Sciences also evaluated numerous claims of "remote viewing." This is the extrasensory ability to describe a location without ever seeing it. Their report concluded that "the case for remote viewing is not just very weak but virtually nonexistent."[13]

These negative results by no means disprove the existence of ESP. But they do demonstrate that evidence in support of the existence of ESP is unacceptable. As parapsychology investigator Susan Blackmore notes, ESP phenomena has been studied for more than a hundred years without anyone turning up repeatable evidence of its existence.[14]

PSYCHICS

A number of people known as psychics have earned reputations for their startling accuracy in predicting the future. Psychics differ from other diviners of the future in that they do not use a detailed system of observation to deduce the future. Rather, they claim to possess a special extrasensory gift that reveals the future to them.

One of the most famous psychics in history was the French physician Michel de Nostredame, commonly known by the Latin name Nostradamus. In 1555, Nostradamus published ten volumes of mysterious four-line verses of prophecy. One of these spoke of a young lion killing an old lion on the field of battle by putting out the elder's eyes in a cage of gold. Four years after the book came out, King Henry II of France was jousting against a younger opponent when one of the com-

Michel de Nostredame (1503–1566), known as Nostradamus, was a French physician and astrologer. His books of prophesies still attract a wide audience.

battants' lances shattered. A splinter of the lance went through an opening in the king's golden visor and pierced him in the face. The king died of the wound, thus assuring Nostradamus of acclaim as a predictor.[15] In recent times, Nostradamus has been credited with predicting the rise of Adolf Hitler in Germany and the Japanese offensive in the Pacific in 1941.[16]

Today, the competition to claim the mantle of Nostradamus is fierce. Television advertisements commonly offer 900 numbers to entice viewers into paying money to complete strangers in order to hear what the future holds in store for them. One of the most famous modern psychics is Jeane Dixon, who also works as an astrologer. Her main claim to fame was the following prediction, made in 1956: "As for the 1960 election, it will be dominated by labor and won by a Democrat. But he will be assassinated in office."[17] That is exactly what happened to Democratic party nominee John Kennedy.

Future-divining psychics claim that their information comes from powers beyond the five senses. Since science deals only with information that can be obtained through the senses, one might expect that a discussion of psychics lies outside the realm of science. But psychics gain acceptance by presenting evidence that their predictions come true. This evidence can be evaluated by our senses to tell us whether or not predictions come true, and it is worth running the evidence through the scientific criteria to see how well predictions hold up.

Psychic predictions, it turns out, are frequently defined in general or obscure terms, in violation of scientific criterion #2. Nostradamus's predictions were phrased in such vague, mystical terms, mixing several languages, that a hundred people could walk away from them with a hundred different interpretations. Anyone wishing to prove Nostradamus was psychic can easily do so by selectively interpreting his words. In reality, King Henry's visor was not made of gold, nor was he struck in the eye, nor did the accident occur on the field of battle, nor was his jousting opponent considerably younger.[18] Yet a loose interpretation can make it appear as though Nostradamus predicted the entire incident. Jeane Dixon predicts such things as

Jeane Dixon, author and self-proclaimed clairvoyant.

Whitney Houston's greatest problem in 1986 being "balancing her personal life against her career." She said that 1987 would be a year "filled with changes for Caroline Kennedy."[19] Both predictions include vague phrases that could be taken to include a vast range of possibilities.

When psychics run the risk of specific predictions they often claim credit for their successful predictions while glossing over their more frequent inaccurate ones. This violates the scientific criterion of objectively evaluating all the data (criterion #8). Jeane Dixon gained fame for predicting John Kennedy's election and assassination. Actually, she also said that Kennedy would *not* be elected.[20] Her accurate prediction was the more vague claim that a Democrat would be elected in 1960 and assassinated in office. Among Dixon's many misses that surrounded the Kennedy "hit" were a prediction of war in China in 1951, a Soviet invasion of Iran in 1953, and the assurance that the Soviets would be first to put an astronaut on the moon.[21]

In burying the evidence of their failures, psychics are helped by the media. Wrong predictions are not news. Anyone can make wrong predictions—it happens every day. Correct predictions are fascinating and make good news stories. As a result, correct predictions get noticed and incorrect ones ignored. This obscures the effect of the laws of chance that say in a given number of predictions (especially vague and trivial ones) a certain number are expected to come true.

In 1979 a man named Richard Newton went much farther out on a limb than most psychics in issuing a detailed prediction. Newton claimed that on March 15, 1980, an airplane with a red logo on its tail would crash just outside a major population center in the Northern Hemisphere. Forty-five people would die.

On March 17, a Royal Jordanian Airlines jet with a red logo on its tail crashed outside Doha, Qatar, killing forty-three people. Such a precise prediction could not have been luck; Newton had apparently demonstrated astounding psychic abilities.

Newton, however, claimed no psychic abilities. He simply made a very psychic-sounding prediction based on a study of history and statistics connected with airplane disasters. He found out that airlines commonly had red logos, that they usually crashed near major metropolitan areas in the Northern Hemisphere, that the average death toll was forty-five, and that March was historically the most common month for airline crashes.[22]

NEAR-DEATH EXPERIENCES

In July 1918, author Ernest Hemingway was seriously wounded by a piece of shrapnel while fighting on the American lines in World War I. Hemingway later reported that, after being wounded, he felt his soul being drawn out of his body, "like pulling a silk hanky out of a pocket by one corner."[23]

What Hemingway experienced is called an out-of-body experience—the distinct feeling that a person is detached from his or her body. One of the most commonly reported forms of out-of-body experience is the near-death experience. One in twenty Americans claims to have had one.[24]

Descriptions of near-death experiences tend to be remarkably similar. Most people talk about feeling a strange peace and becoming separated from their body. They enter a darkness or tunnel. As they move forward through the darkness, they see light in the distance, and then enter the light. Often these "journeys" are accompanied by music and flashbacks. At some point, the travelers reach a barrier and realize that they must go back. They return to their body filled with an indescribable feeling of peace and joy.[25]

Some people cite near-death experiences as evidence that there is life after death. Others believe these reports are nothing more than hallucinations caused by lack of oxygen to the brain. These critics point out that all the near-death-experience elements have also been described by people who did not have a near-death experience.[26] Are claims of near-death experience, then, a pseudoscientific fraud?

Near-death experiences defy the scientific criterion of falsifiable hypotheses (criterion #3). There is simply no practical way of scientifically testing a near-death experience. No respectable researcher would think of inducing a state of near death in a patient for purposes of experimentation. Neither are emergency victims who lie near death proper subjects for casual experimentation. Further, even if science could determine a chemical or biological link with the thoughts that occur in a near-death experience, science does not provide a basis for interpreting what those thoughts mean.

Therefore, neither the claim that near-death experiences give evidence of afterlife, nor the claim that they do not, can be scientifically valid.

WARNING SIGNS OF PSEUDOSCIENCE IN PARAPSYCHOLOGY

The above are just a sample of the world of parapsychology. The various forms of ESP and all the evidence for and against their validity are too numerous to detail. But special attention can be paid to some of the pseudoscientific faults that parapsychology is most apt to fall into.

James Alcock defines parapsychology as "the attempt to study the independence of the mind from the world of the physical."[27] Since parapsychology is the study of things that are not logical, criterion #1 about logic and common sense cannot automatically be held against it.

Claims that violate the rule about falsifiable hypotheses (criterion #3), however, carry a strong scent of pseudoscience. For example, scientific experiments have been unable to find any evidence that telepathy, the ability to communicate from mind to mind, exists. Some of those who believe in telepathy have explained this by saying that "science cannot be expected to detect telepathy since unsympathetic skepticism inhibits it."[28] In other words, even negative experimental results are automatically explained in a way that supports the existence of telepathy.

When world-famous psychic Uri Geller was unable to demonstrate any psychic ability to a group of scientific observers, his response was, "I am not a radio or TV that you turn on or off."[29] When a parapsychologist was caught manufacturing data on an experiment, one of his supporters explained that it was simply an example of the man's "powers of precognition" that he was able to know what the results would be ahead of time.[30]

All of these are classic nonfalsifiable hypotheses. They can explain all possible data, even contradictory data.

Some "scientists" have claimed the discovery of a new type of ray called the E ray. These are said to come from deep in the earth and can cause health problems. Unfortunately, they are not detectable by any scientific instrument, only by a willow dowsing stick. This nonfalsifiable hypothesis should send warning signals to any scientist. If E rays cannot be detected by science, how can they be proved or disproved? But instead of heeding this pseudoscientific warning, the German government paid a quarter of a million dollars in 1990 to hire dowsers to scan hospitals and federal offices for E rays.[31] This controversy mirrored that surrounding René Blondlot's claim to have discovered "N rays" in the early part of the century (see chapter 5).

Claims in support of parapsychology frequently violate the scientific criterion of impartial examination of evidence (criterion #8). A lengthy study of ESP by a Soviet research team concluded, "In parapsychology, faith is more aggressive than facts."[32]

Uri Geller displays a fork he allegedly bent using his psychic powers.

A report by skeptic Samuel Gill demonstrates this vividly. Gill infiltrated a group that was convinced of the extrasensory powers of their leader. At one point in the meeting, the leader passed around an invisible "healing ball of light." Each person was to handle this invisible ball, absorb its powers, and pass it on to the next person. Afterward, members of the group enthusiastically described the benefits they had gained from the invisible ball. Gill then revealed that when he had received the ball, he kept it in his fist and not passed it around! Yet those who had not received the "ball" claimed the same benefits as those who had.[33] A mind that is strongly inclined to believe a certain way cannot be a reliable judge of scientific evidence.

The scientific criterion of peer review (criterion #5) has recently been violated by a number of psychics who have resorted to legal tactics to avoid having their claims tested in open debate. Magician James Randi has made a career of exposing psychic frauds and the errors of pseudoscientists. He has had to spend hundreds of thousands of dollars defending himself against libel suits. The threat of libel action discourages most investigators from examining the claims of psychics. As James Grossberg, lawyer for the Committee for the Scientific Investigation of the Claim of the Paranormal (CSICOP), notes, "if people can't comment on claims of miracles, we're in real trouble."[34]

Another indicator that crops up when evaluating claims of parapsychology is caution in examining the evidence (criterion #7). The history of the paranormal is filled with impressive displays that turned out to be hoaxes. One of the best known is the Amityville case, which was popularized by the motion picture industry. In 1974 six family members were murdered by a seventh family member in their Amityville, New York, house. The following year the house was bought by George and Kathy Lutz. What followed was twenty-eight days of terror as the couple was tormented by weird events such as green slime oozing from walls, strange voices, and doors being ripped from their hinges.

After an enormous amount of publicity, though, the story turned out to be made up for money. In a 1979 interview William Weber, who was writing a book on the original murder, admitted that he and the Lutzes "created this horror story over many bottles of wine."[35]

In April 1981, Cable News Network presented excerpts from a January 6, 1981, taping of a prediction by psychic Tamara Rand. Rand, who made a living advising movie actors and actresses whether or not to sign contracts, predicted on a talk show that President Ronald Reagan would be shot in the chest in late March by a young man with the initials "J.H."—something like "Jack Humley." On March 30, Reagan was indeed shot in the chest by John Hinckley. An amazing demonstration of psychic power? Four days later the talk show host confessed that the prediction was actually taped the day *after* the shooting.[36]

G. S. Soal achieved startling results in experiments demonstrating the existence of mental telepathy in the 1940s. Later it was discovered that he had gained those results by manipulating the data.[37]

Harry Houdini, perhaps the most famous magician of all time, was obsessed with a desire to communicate with his dead mother. During his life, he found a great many psychics who claimed to be able to help him do just that. As a master illusionist, however, Houdini could not be fooled. In every case, he uncovered the fraud.[38]

Modern-day magician James Randi is similarly tough to fool. He has repeatedly exposed apparent displays of psychic power by revealing the tricks behind them. He insists that many of these displays are "very simple tricks that kids have been doing for ages."[39] One of these was the ability of psychics to bend a spoon simply by the force of their thought. Randi has personally performed the spoon-bending trick on many occasions. He carries a $10,000 check with him at all times, payable to any person or group who can perform one paranormal feat of any kind. He has never paid.

One of Europe's most well-known psychics, Hans Bender, begins one of his recent books with the story of an American college student named Lee Friend. In 1977 Friend had a dream in which a friend who had recently died showed him a newspaper. The headlines told of a collision of two 747 aircraft over Tenerife that left 583 dead. Friend related his dream to his college president. Ten days later the disaster occurred just as he had foretold. According to Bender this case was "reliably documented." In fact it has been firmly established as a hoax. Before the crash took place, Friend had put an unrevealed prediction in an envelope. This envelope was then placed in a locked drawer in the president's office. After the disaster, the drawer was unlocked and the prediction of the 747 crash was found. Friend, however, admitted that he had played a trick by secretly inserting his prediction into the drawer *after* the disaster had been reported.[40]

There are honest people who seriously believe in and conduct research into paranormal phenomena. But the history of parapsychology shows a disturbing frequency of fraud. Further, the very nature of the subject has enormous potential for fraud. According to one who has used them, magic tricks can be so sophisticated that not even the most stringent scientific control will detect them. On top of that paranormal subjects offer a lucrative temptation for committing fraud. "Belief in the paranormal is fed and reinforced by a vast media industry that profits from it."[41] All of these factors taken together should make anyone leery of jumping to conclusions about its validity.

The claims of paranormal advocates are extraordinary, and one of the elements of scientific objectivity is that extraordinary claims require extraordinary proof. Police in many localities regularly consult psychics to help them locate clues, criminals, and murder victims. Is there extraordinary proof to show this is a good use of time and money? On the contrary, despite the hype by the press, police psychics have been found to be not only useless but harmful to the investigations.[42] As an example, police in one case issued an open invitation to psychics to help them find a missing person. They received nearly 1,200 responses from those who claimed extrasensory powers. As a policeman described it, "We listened to them all, but they didn't do a thing to help our inquiry."[43]

Dowsing ability is an extrasensory gift that has gained significant acceptance, as evidenced by the more than 25,000 practicing dowsers in the United States.[44] Dowsing involves locating a hidden object, person, or substance with an instrument such as a willow stick that, through some unexplained method, will point in the direction of the desired object. Most dowsers use their skill to locate underground water.

Where is the extraordinary evidence that this extraordinary ability exists? The National Research Council of the United States summed up the situation for dowsing as well as other paranormal claims when it reported in 1988 that it found "no scientific justification from research conducted over a period of 130 years for the existence of paranormal phenomenon."[45]

Finally, suspicion is in order whenever paranormal claims offer coincidence as proof (criterion #9). After hearing claims that weird things began happening to viewers while a psychic was appearing on a television show, James Randi conducted his own experiment to see if there was a true connection between the psychic and the strange happenings, or whether the happenings were a normal result of the law of averages combined with active imaginations.

Randi arranged for a psychic to appear as a guest on a radio show. He then invited listeners to phone with reports of strange things that were occurring in their homes in response to this psychic energy. Sure enough, the phones started ringing with tales of light bulbs exploding, mirrors shattering, and toilet paper falling off rolls. Randi then revealed that his "psychic" was an imposter. His point was that at any given time, abnormal events occur.[46]

One of the most commonly cited evidences of paranormal behavior is the dream that comes true. What else could that be but ESP? Sleep researchers, however, point out that the average person has five periods of dreaming each night, with about fifty dreams occurring during each period.[47] Multiply that by 5 billion people in the world and you have more than a trillion dreams in the world every night. Even if dreams are just random splicings of unconscious thought, the odds are pretty good that something out of those trillion dreams will resemble reality. A valid scientific test would require a great many controls, including keeping a diary in which dreams are written down to prevent them from being misremembered.

Suppose a dream really does come true. The coincidence is just too incredible to attribute to chance. How does science account for that?

Science does not pretend to account for everything that goes on in the world, for every rare instance that defies explanation. Stories abound of premonitions of disaster, miracles of healing, wondrous escapes against all odds, and sudden, radical changes of heart that defy logical explanation. A Soviet report on the subject of parapsychology concluded that "some so-called paranormal" phenomena do exist. But, as the report also concludes, in words echoing the report of the National Academy of Sciences, "so far no physical basis has been demonstrated for it."[48]

It remains to be seen how much use, if any, science can be in helping organize the mysteries and bizarre quirks of nature in some systematic and understandable way. Alcock states flatly that "parapsychology is indistinguishable from pseudoscience and its ideas are essentially those of magic."[49] While that may be too strong a charge, until parapsychologists can offer solid proof to the contrary, only caution can prevent pseudoscience from exploiting the unanswered questions of our existence.

FOUR

BATTLEGROUND: THE HUMAN BODY

In 1858 a miracle medicine hit the market in England. One advertisement for this mysterious potion promised that it would restore "perfect digestion, strong nerves, sound lungs, a healthy liver, refreshing sleep, and energy" and would cure "colds, flu, headaches, hemorrhoids, asthma, and dozens more conditions."[1]

This product was just one of many patented medicines that flooded the market at that time. None of them did what they claimed, but they earned fortunes for the manufacturers at the expense of gullible consumers. Fortunately today's consumers have come a long way from those days of ignorance. Or have we?

Arthritis is a painful stiffening of the joints that affects 40 million Americans. According to the United States Food and Drug Administration (FDA), 95 percent of arthritis sufferers use some form of self-treatment[2] ranging from copper bracelets and Chinese herbal remedies to snake and bee venom, alfalfa and algae pills, and even cow-dung baths. "There are so many phony arthritis cures," says one official, "that we can't keep track of them."[3]

"Health fraud is an extraordinarily lucrative business and it is probably growing," according to William

Jarvis, president of the National Council Against Health Fraud.[4] Estimates of the amount of money that Americans waste on quack health products or treatment range from $27 billion to $40 billion dollars per year. The FDA estimates that 38 million Americans use bogus health products each year, and that one of ten is actually harmed by them.[5]

The government can do only so much to prevent this thievery. According to Ken Durham of the FDA, "Health fraud is bigger than any one organization can deal with."[6] The buyer often has no help in trying to decide whether a health product or service is legitimate. Unfortunately, spotting health fraud has never been more difficult for the average person. Medical research is surging ahead at such a rate that no one can keep up with all the advances. Profit-hungry merchants have taken advantage of the information gap.

"Quacks are more sophisticated today," explains William Jarvis. "They wear the white coat of science."[7]

The scope of medical pseudoscience is far too vast to cover here in its entirety. But the science criteria can help to distinguish a legitimate health care claim from a phony one.

MIRACLE CURES

Scientific research is a careful, cautious process. While legal pseudoscientists latch onto fragments of data to persuade juries that impact with a steering wheel causes lung cancer, true scientists require a thorough investigation before they would accept such a conclusion (criterion #4).[8]

E. Cuyler Hammond, chief epidemiologist at the American Cancer Society, spent more than thirty years probing a possible connection between smoking and lung cancer. His study, published in 1954, examined nearly 188,000 cases. It was one of the first reports to provide strong evidence that smokers were twenty times more likely to develop lung cancer than nonsmokers. This study was followed by others that confirmed Hammond's findings. As a result of this long, gradual process, the link between smoking and lung cancer was firmly established in medical circles.[9]

People who contract deadly diseases such as cancer are understandably impatient with the painstaking process of medical research. They need a cure and they need it now. They can't wait for results. Advances in medical science occur frequently enough to continually hold out the hope that a miracle breakthrough is on the horizon. Seriously ill patients may try anything that sounds the least bit plausible and gives the tiniest glimmer of hope.

Frustration with a costly and inefficient medical system further steers patients away from established medical science, especially when they feel their doctors are uncaring or inattentive to their needs. Impatience, desperation, and frustration are ideal breeding grounds for pseudoscience. Not only can fraudulent cures be expensive, they can kill a patient who chooses them in place of legitimate treatment.

The standard, established method of fighting cancer involves surgery to remove cancerous tissue, and radiation and chemotherapy to kill cancer cells. Recently, doctors have begun to emphasize additional treatment intended to improve the patient's overall health. These techniques became accepted after years of study and testing. The effectiveness of generally recognized methods can be seen in the improving survival rates of cancer patients. Currently, half of all diagnosed cancer cases are cured,[10] including most cases of childhood leukemia, Hodgkin's disease, and testicular cancer, and many breast, colon, bladder, and skin cancers.[11]

For those who want more than standard health care

can provide, alternative cancer treatments are available. But are these treatments legitimate? Have they been proven effective or are they phony cures hiding behind a screen of pseudoscientific evidence? These alternative treatments include:[12]

1. Nutritional programs and diets. Research has established that a low-fat, high-fiber diet can help prevent cancer. This has led some to the conclusion that certain diets can actually cure cancer.
2. Therapies designed to improve the cancer patient's immune system. Some of these therapies use injected vaccines to stimulate production of disease-fighting agents in the bloodstream. One immune-stimulating agent called Levamisole was approved by the National Cancer Institute in 1989 for use in combination with standard chemotherapy treatment. Some researchers claim that dilute blood plasma can provide a boost to the immune system.
3. Metabolic treatments. These involve introducing a variety of cancer-fighting substances, such as vitamins, minerals, and enzymes. Laetrile, found in apricot kernels, has been one of the more famous products tried. Some of these treatments are based on the theory that certain toxins in the body interfere with healing, and that better health can be achieved by ingesting something that neutralizes these toxins.

Is there any way for a concerned person to recognize whether a treatment is legitimate? All of the above are based on medical principles and employ scientific terminology. But that is true of most pseudoscience. The scientific trappings are what distinguish pseudoscience from superstition and folklore.

Woman undergoing crystal healing.

We must look to the science criteria for guidance in separating legitimate science from the bogus. Medical pseudoscience will most often reveal itself by failing to meet four scientific criteria.

One of these is caution in examining the evidence (criterion #7). Bold headlines that proclaim miracle cures in sensational language are warning signals that lead us to ask what purpose this sensational hype serves. Sensational claims can help the media attract readers. They can bring publicity that might be useful to a research group trying to attract funds for further research. They can help persuade gullible people to buy the product. One thing that sensational claims cannot do is promote the truth. In the scientific world, only evidence can do that. Any account of a miracle cure that is long on glowing praise and short on facts is suspect.

Of course, science does occasionally produce breakthroughs that seem miraculous. But good scientists do not need hype to persuade or inform people; they can do it with facts.

In 1985 three French scientists informed reporters at a news conference that the drug cyclosporine appeared to halt growth of the AIDS virus. By anyone's standards that was a tremendous medical breakthrough. Such a momentous claim should not be accepted without a careful look at the facts.

At it happened, the claim was based on the observation of two AIDS patients for just over a week. Anyone familiar with scientific research would know that this was a ridiculously small amount of data on which to base any firm conclusion. At best, this could have been called a promising lead. The media, however, immediately trumpeted this "medical breakthrough" as front-page news. Within days of the story, one of the two AIDS patients died. So much for the miracle cure.[13]

In this case the phony cure was easy to detect. But many "cures" peddled for a variety of illnesses are not. They are promoted as if they really are the result of scientific research, and purchased by people who accept that at face value.

A second scientific criterion that helps reveal medical pseudoscience is the matter of allowing claims to be examined by peers (criterion #5). Any "secret" cure is automatically suspect. Unless the cure is laid out in the open so that other knowledgeable people can examine the evidence and perform their own tests, it cannot be taken seriously. Miracle cures that have not been described in any scientific journal or are not backed by extensive data must be viewed with suspicion. Where are the studies showing that Laetrile can cure cancer or that certain diets can strengthen the immune system? If those studies are not available, chances are the research has not been done or the quality has been too poor to gain acceptance from scientific peers, or the studies showed that the cure did not work.[14]

Many medical claims rely primarily on anecdotal evidence for proof, a direct violation of criterion #10. A number of patients will swear that they tried a certain treatment and, to the befuddlement of their doctors, their cancer was cured.

Science cannot accept this as convincing evidence because these reports are made by nonscientists who, for the most part, have no understanding of the concept of experimental controls. The fact that a certain therapy was tried and a person recovered does not, in the eyes of science, prove that the therapy worked. The fact that an unusual number of patients undergoing a certain therapy improved does not prove that the therapy worked. Too many undetermined variables exist. Scientists need to know whether there was anything unusual about that group that might account for the result.

Testimonials from cured patients also run into problems with the science criterion concerning coincidence (criterion #9). Suppose a patient tries a novel arthritis treatment and immediately the pain goes away. The patient and the patient's family could likely conclude that the treatment worked. A careful medical researcher, however, knows that the symptoms of arthritis go into remission from time to time without treatment.[15] By the laws of chance, many arthritis sufferers will try a remedy at the time their ailment goes into remission and will falsely connect the two.

Suppose a terminal cancer patient tries an exotic cure and then lives five times longer than expected. The person could hardly help but rave about the miracle cure. A careful medical researcher, however, realizes that a certain number of cancer victims survive well beyond the expected. There are even rare cases in which terminal cancer patients recover completely for no apparent reason.[16] This level of coincidence must be taken into account in evaluating testimonials.

The development of a drug called taxol has recently attracted a great deal of attention as a cancer cure. Taxol is an extremely complex chemical found in the bark of the rare Pacific yew tree. Recently, chemists have discovered how to synthesize taxol from less scarce sources.

The effect of taxol was discovered in the early 1960s during a National Cancer Institute (NCI) search for plants that showed anticancer potential.[17] Early reports of its cancer-curing ability were confirmed in tests at Johns Hopkins Medical School. In 1979 a team led by Susan Horwitz at the Albert Einstein College of Medicine discovered how taxol works; it "freezes" cancer cells so they are unable to divide and grow.

Repeatable tests performed by the NCI showed that taxol increased the survival time of mice with cancers. Further tests have shown that 30 to 40 percent of patients with advanced ovarian cancer respond to taxol treatment. In more recent experiments, taxol has shrunk tumors in about half of the breast cancer patients tested, and even caused the cancer to disappear in some cases.

Researchers concede there are problems with the drug. They point out that even in patients who respond to treatment, the tumors return as soon as the treatment is stopped. Present means of producing taxol are far too expensive for it to be practical. Several patients have died from side effects that include lowered blood cell count. As Zola Horovitz, a researcher for the Bristol-Meyers company that is working to develop a taxol product says, "Taxol is a step forward, but it's not a miracle drug and it's not a cure for cancer."[18]

Can taxol be trusted? In this case, there are no obvious warning signs in the form of sensational, unsupported claims, limited data, absence of peer review, or reliance on coincidence and anecdotes for proof. Therefore there is no reason to doubt taxol shows promise as a cancer-fighting agent.

Vitamin C has been touted as a multipurpose remedy capable of curing everything from colds to cancer. Its benefits have been championed by Nobel Prize–winning chemist Linus Pauling. Pauling advocates enormous doses (300 times the generally accepted standard) and cites studies that show vitamin C prevents colds and prolongs the lives of cancer patients.[19]

Other scientists scoff that large doses of vitamins give you nothing but "expensive urine."[20] They cite experiments showing that vitamin C has no effect on colds or cancer. Each group criticizes the other's experiments as being faulty or poorly designed.

Is one side or the other making an honest mistake or is pseudoscience involved here? The indicators show that there may be some of both. Many advocates and critics of vitamin C follow the rules of science. Others do not. Vitamin C supporters often fall short of scientific standards by basing their conclusions on skimpy evidence (criterion #7). As *Science* magazine comments about vitamin C boasts, "claims like that would make any cautious researcher run for cover."[21] A spokesman for the FDA agrees, saying, "Somebody has made practically every claim you could dream of

A drugstore vitamin display. It's often hard to separate substantiated claims about vitamins from hearsay.

about these vitamins. You never know what next year's fad is going to be."[22]

On the other hand, according to another science researcher, "scientists are starting to suspect the traditional medical views of vitamins and minerals are too limited."[23] Legitimate researchers are beginning to compile evidence that meets scientific standards. Some recent evidence has shown that eating foods high in vitamin C lowers risk of cancer, and that the vitamin may reduce some of the side effects of cancer treatment. Scientists are exploring the intriguing possibility that vitamin C works as an antioxidant to neutralize harmful chemicals in the body before they can cause trouble.[24]

Conflicting evidence suggests that scientists have only scratched the surface of vitamin research. At the rate that new information is being uncovered, scientists may soon unravel the mysteries of how vitamins affect health. As with any new frontier science, the scientific criteria warn us away from exaggerated claims while keeping an open mind to new evidence that is being gathered.

MIND AND BODY MEDICINE

As we saw in the preceding chapter, the human mind is such a bewilderingly complex organ that any scientist trying to probe its depths can get lost in a hopeless quagmire. The problem extends to medical researchers trying to get a handle on how the mental processes affect physical health.

For many years Western medical practitioners have looked down their noses at ancient traditions that claimed to heal the body by healing the mind. The 3,000-year-old Chinese art of acupuncture and the Navajo tradition of healing by restoring the patient to balance with nature have long been viewed as remnants of primitive superstition.

In recent years, there has been a renewed interest in ways that the mind can help heal the body. Acupuncture, hypnosis, biofeedback, meditation, yoga, psychological support groups, and crystal healing have all enjoyed a surge in popularity as methods of healing. These techniques are attractive because they are pain-

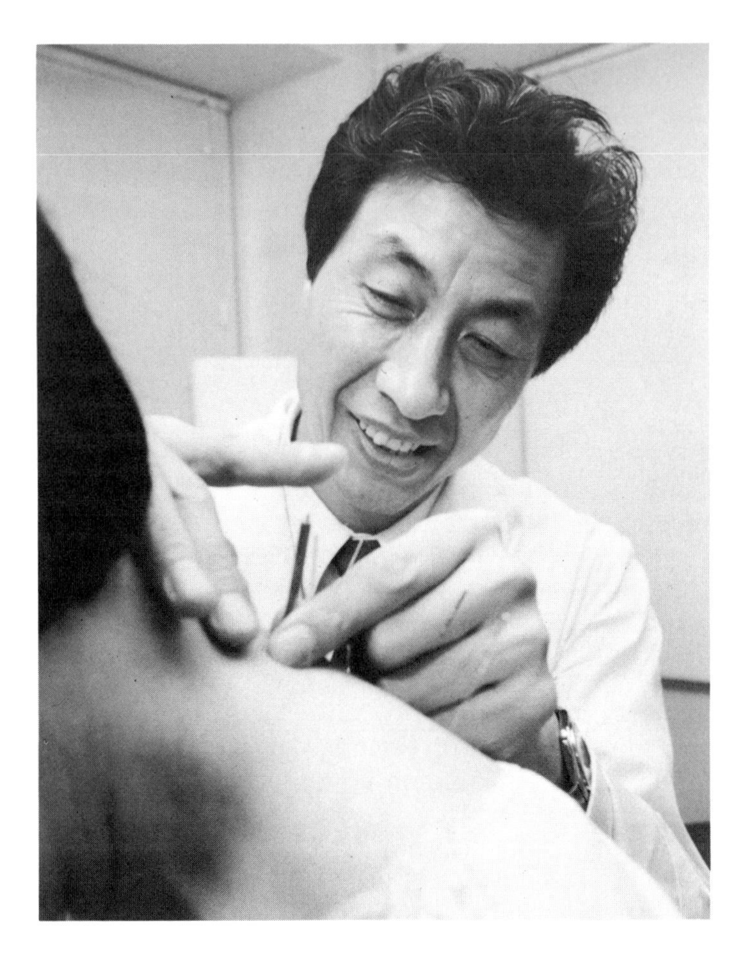

A Chinese acupuncture specialist practices his art.

less, less expensive, and more user-friendly than more traditional medical methods such as drugs and surgery.

But do they work, or are they pseudoscience? Even the scientific community has had trouble with this. Physician Dean Ornish at the University of California at San Francisco states that "psychological measures used to treat or prevent heart disease are just as important as anything else we do."[25] Other researchers and doctors complain that mind-body medicine consists of far more hype than evidence.

Acupuncture is a technique for relieving pain that involves inserting tiny needles into the skin at precise points. The hairlike needles, which are too small to cause pain, are left in place for anywhere from ten minutes to an hour. Acupuncture was discounted by Western doctors and virtually unknown in the United States until the early 1970s. At that time respected political columnist James Reston wrote about the benefits he received from acupuncture while on assignment in China.[26]

Reston failed to convince the American medical community. His report, after all, was merely anecdotal evidence (criterion #10). Acupuncture seemed to have all the earmarks of a classic pseudoscience. The Chinese explained the success of acupuncture by claiming that the points on the skin are connected to certain organs and bodily functions. An acupuncture point near the wrist, for example, is associated with respiration. Pain is caused by the interruption of the normal flow of bioelectricity between points on the body. Stimulation of the proper connection with needles restores the proper flow.[27]

The explanation made no sense according to Western thought. It was not logical (criterion #1) because research into the systems of the human body did not provide a basis for the existence of such a bioelectrical system. It was nonfalsifiable (criterion #3): How could you prove that acupuncture did not restore bioelectricity when there was no evidence that any such thing existed?

Less skeptical consumers, though, especially those desperate for relief from chronic pain, gave it a try and reported positive results. At the same time, studies demonstrated that acupuncture could indeed relieve chronic pain.

Since the data were conflicting with some of the science criteria, unbiased observers could not be sure what to make of acupuncture. Finally the deadlock was broken by research that showed that the needles caused the brain to release substances called endorphins. Endorphins provide the logical mechanism by which acupuncture relieves pain. These substances are capable of slowing the response of neurons to painful stimuli.

Acupuncture has not totally shed its old pseudoscientific image. Health insurance companies consider the techniques unproven and rarely cover costs. The majority of acupuncture practitioners in the United States are not trained physicians. Out of 500,000 physicians in the United States today, fewer than 3,000 use acupuncture in any way.[28]

Why the brain produces endorphin in response to needles remains a mystery. But now that scientists have some kind of reasonable explanation for how acupuncture works to go along with evidence that it does, acupuncture can be taken seriously. It has been shown to be effective for patients with chronic back problems and arthritis. A hospital in New York City reported that acupuncture was so successful at easing the pain of withdrawal from alcohol and drugs that an exceptionally high percentage of its hard-core addicts remained clean after three months.[29]

Hypnosis is a way of inducing a state similar to sleep, during which the mind is susceptible to certain suggestion. As a common butt of hocus-pocus satire, hypnosis seems a likely candidate for pseudoscience. But in fact its use as a mind-body therapy has been accepted by the American Medical Association for nearly forty years.[30] More than 15,000 physicians supplement conventional treatment with hypnosis to treat patients with arthritis, migraine headaches, and even burns. Studies have found that cancer patients who are hypnotized before undergoing chemotherapy are less likely to experience nausea. Burn patients treated with hypnosis recover more quickly, with less medication. A ten-year study at Stanford University found evidence that breast cancer patients who used self-hypnosis to ease discomfort survived longer than those who did not.

As with acupuncture, no one is sure why hypnosis works. The therapy does not appear to have any direct effect on the medical condition; rather it seems to help unleash the body's natural ability to heal. Again this delves into a mind-body connection that remains a mystery. But evidence for its success appears in mainstream medical journals and the presentations do not violate any obvious science criteria. Former skeptic Karen Olness of Case Western Reserve University in Cleveland describes how hypnosis distances itself from pseudoscience. In the medical world, she says, there is a "lot of lip service about mind-body connections. Hypnotherapy provides a tool to document it."[31]

Biofeedback is a therapy in which people mentally control conditions that are not normally thought of as controllable, such as heart rate and pain. The patient concentrates on controlling the function or condition, usually by trying to relax muscles. Sensors attached to the body tell the patient how well his or her concentration is working. Again, no one knows why it works, but biofeedback has shown success in relieving pain, stress, and migraine headaches, and shows promise in other areas as well.[32]

Meditation also offers promise. An Ohio State University study has measured a slight increase in the strength of the immune systems of retirement home people who meditated.[33]

Harvard physician Walter B. Cannon, in his book

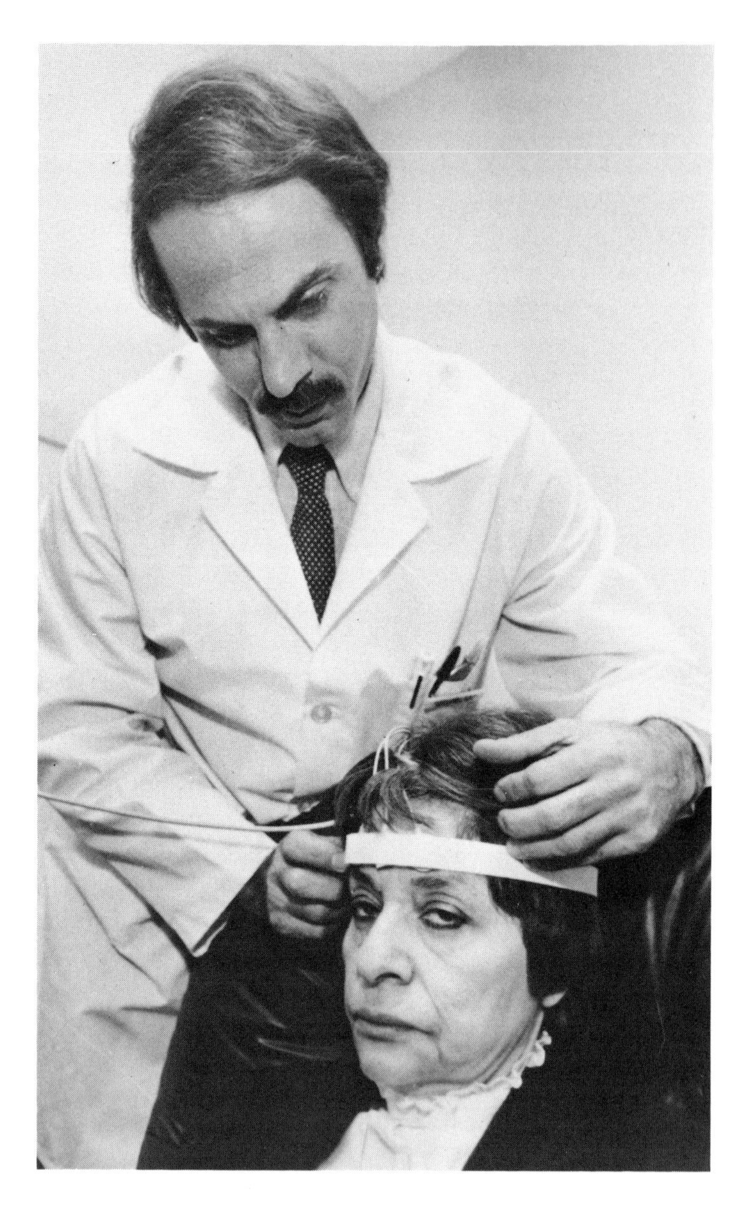

In biofeedback therapy, patients learn to regulate bodily functions. The sensors on this woman's forehead are hooked up to a biofeedback machine that will help her learn to relax her facial muscles to relieve headaches.

Voodoo Death, described the extreme example of the opposite effect of the mind. He documented cases in which a voodoo curse actually caused death in voodoo believers. The victims of the curse went into self-induced shock and suffered circulation failure, breakdown of oxygen intake, and starved vital organs.[34]

If biofeedback, meditation, and voodoo work, how far will people carry the conclusion that they can "think" their way to health? Does this mean that people can beat cancer through sheer willpower? Does it mean that doctors are unnecessary? There are those who make both claims.

When generally sound scientific procedure is combined with gaps in understanding, chances are we are dealing with a new-frontier science. But gaps in understanding can easily be filled in by pseudoscience.

The uncertainty surrounding the mind-body health connections leaves the criterion of caution in examining evidence (criterion #7) as the main buffer zone between patients and pseudoscience. Especially in matters of life and health, we should be extremely careful in examining the evidence and suspicious of those who jump to conclusions.

For example, the question can be raised that if mind-body therapies have been shown to work despite our ignorance of why they work, then why hasn't crystal healing? This therapy holds that the body can obtain healing energy from quartz. Bizarre, maybe, but is it any more mysterious than acupuncture?

According to criterion #7, extraordinary claims must be backed by extraordinary evidence. The claim that rocks emit a healing energy that scientific instruments are unable to detect is so extraordinary that it requires enormous proof to be taken seriously. The fact that this energy evidently cannot be measured also violates the scientific principle of the falsifiable hypothesis (criterion #3). Crystal healing clearly falls into the ranks of pseudoscience.

Voodoo rite in Haiti.

BIOLOGICAL RHYTHMS

The idea that human processes are greatly affected by internal rhythms originated in two different ways at two different times. The first approach was developed by Wilhelm Fliess, a Viennese surgeon of the late nineteenth century, who was a close friend of psychoanalysis pioneer Sigmund Freud. Fliess was a nose and throat specialist who held some interesting medical theories.[35] He believed, for example, that the nose was responsible for many ailments of the nervous system, and he prescribed cocaine application as a cure.

Fliess also came to believe that the numbers 23 and 28 had an almost mystical significance. Out of his thought and study came the idea that the human body follows cycles of efficiency. A person's physical cycle, according to Fliess, lasted twenty-three days; the emotional cycle twenty-eight days. After Fliess's death, his followers added a thirty-three-day intellectual health cycle to the system. These cycles became commonly known as biorhythms.

According to the Fliess system, these fixed cycles begin at the moment of birth. During the first half of the cycle, a person's abilities are on an upswing and during the last half, on a downswing. The midpoints of each cycle—that is, when the pendulum swings from one direction to the other—throw the body off stride. On those "critical days," a person functions at the lowest level of efficiency and is vulnerable to accidents and poor performances. The most vulnerable day of all is the "triple critical" day, the once-a-year occurrence when all three cycles hit the critical midpoint. Based on this system, up and down days can be predicted based on a person's birthday.

In the 1930s, Swiss researcher Hans Schwing matched biorhythms against 1,400 accidents. His analysis showed that 60 percent of the accidents, and 65 percent of the fatal accidents, occurred on critical days.[36] Since critical days occur only 20 percent of the time, this indicated a tremendous impact.

The biorhythm theory attracted a small following in Europe after World War II, and was introduced to the United States in the 1960s by Swiss-born industrial consultant George Thommen.[37] By the 1970s, biorhythm was a mushrooming business, serving more than a million Americans. Supporters included Dallas Cowboys talent scout Gil Brandt, who agreed that the practice "has a lot of validity."[38] Jacob Sanhein of the Naval Weapons Support Center in Crane, Indiana, presented research to the National Safety Council in 1975 in support of biorhythms. His findings showed that 40 percent of accidents occurred on critical days, twice the level predicted by chance.[39] Other studies concluded that from 40 to 80 percent of accidents happened on critical days. A Japanese bus company weighed in with startling evidence that it had cut its accident rate in half by issuing safety warnings on drivers' critical days.[40]

Fascinating historical and personal evidence turned up in support of biorhythms. Custer and Napoleon suffered their great defeats on critical days. Muhammad Ali lost his title on a double critical day.

Facing page: *Biological rhythm theory postulates the phenomenon of "critical days." The football game between the Minnesota Vikings and the San Francisco '49ers that took place on October 25, 1964, supposedly occurred on a "critical day" for Jim Marshall of the Vikings. Marshall picked up a fumble (above) and ran 60 yards in the wrong direction to give San Francisco a safety (below left). Bruce Bosley of the '49ers then congratulated and thanked Marshall in the end zone (below right). Despite this, the '49ers still lost the game, as a result of their own goof-ups.*

70

77

Football player Jim Marshall's famous wrong-way touchdown run in 1964 came on a triple critical day.

The evidence was persuasive enough that management at an Exxon chemical plant sent out safety reminders to their employees on triple critical days. United Airlines began checking into biorhythms for their pilots. One biorhythm proponent predicted, "We are only five years from advertising tag lines like, "Our pilots never fly on critical days."[41]

Some mainstream watchdogs of the science world took cautious stands lending support to biorhythms. *Popular Mechanics* touted a "biolator calculator" for helping people calculate their critical days.[42] In 1973 *Science Digest* published an article suggesting that there might be something to this new phenomenon. The article advised readers, "Don't panic, but be careful" on a critical day.[43]

Those who jumped on the biorhythm bandwagon failed to notice a rash of pseudoscience indicators. The existence of biorhythms had little do with logic (criteria #1). None of the other known biological cycles (sleep, menstrual) had anything to do with birth dates. Nor was there any logical reason why they should. A fetus spent nine months developing in the womb. What was so magical about drawing that first breath of air that would cause the cycles to kick in?

Further, there were unexplained gaps in the theory (criterion #6). Where did these cycle numbers come from? Dr. Fleiss seemed prone to strange, unsupported theories. Upon what data did he derive the twenty-three- and twenty-eight-day cycles?

The stories of Custer, Napoleon, and Marshall, eerie as they might seem, offered no conclusive proof according to the principle that science does not accept coincidence as proof (criterion #9). Out of a hundred disastrous historical events, the law of averages dictates that twenty would be expected to fall on a critical day. Out of a thousand disasters, two hundred would be expected on critical days. Making a list of those disasters proves nothing. A similar list could be made for any day of the twenty-eight-day cycle.

The fact that people were making millions of dollars providing the public with biorhythm expertise called into question the impartiality of the advocates (criterion #8).[44] People who have a great deal at stake in the outcome of research cannot be trusted to handle the inquiry objectively.

The studies cited as proof that biorhythms worked contained enough flaws to provoke concerns about jumping to conclusions (criterion #7). Just because a bus company experienced a 50 percent reduction in accidents when it started alerting drivers to critical days does not prove anything about biorhythms. Wouldn't most people drive more carefully if they were told that they were at risk on that day? Too many controls were missing in the research to warrant any firm conclusions.

Further research provided solid evidence that Fliess's biorhythm system was phony. A Johns Hopkins University team studied 205 serious or fatal highway accidents in Maryland and found that no more of them occurred on critical days than on any other day.[45] The United States Tactical Air Command looked into 8,625 flight accidents and found no biorhythmic effect. The Workman's Compensation Board of British Columbia studied 13,000 accidents and found no basis for the biorhythm claims.[46] As the scientific criteria predicted, biorhythms have been discounted by scientists as "a silly numerological scheme that contradicts everything we know about biological rhythms."[47]

What we do know about biological rhythms originated with Franz Halbert.[48] In 1940 Halbert noticed that the white blood cell count in laboratory mice was consistently higher and lower at certain times of day.

Since then, researchers have been charting various body indicators and have discovered a variety of internal rhythms. The most common is a twenty-four-hour pattern of activity and rest, referred to as circadian rhythm. The term "chronobiology" has been coined to distinguish this field of study from the pseudoscientific biorhythms.

Among the findings of chronobiologists are the fact that blood pressure can vary by as much as 20 percent, body temperature by two degrees (usually peaking in the evening), and white blood cell counts by up to 50 percent. Blood pressure, body weight, and cholesterol and hormone levels also rise and fall in a regular pattern. The changes are caused by chemical and hormonal response to changes in temperature, air pressure, and light.[49]

These changes affect your mood, your alertness, and how well you perform various tasks. For most people, short-term memory work and mental tasks such as mathematics appear to be best handled in the morning. Alertness and decision-making skills peak around noon. Long-term memory and small motor skills are best suited to the afternoon. The senses operate at high efficiency in the late afternoon, and sports are best suited to late afternoon and evening.[50]

Psychologist Martin Teicher of Harvard Medical School believes that depression and other forms of mental illness could be connected to the body's internal rhythms. A two-year study found that depressed children show circadian rhythm patterns that are distinctly different from the usual. These children tended to run on cycles of twelve hours rather than twenty-four. After a period of alertness in the morning, their systems would wind down in the afternnon and then rev up again in the evening. Psychologists who study rhythm patterns see hope that such problems as Attention Deficit (Hyperactive) Disorder can be diagnosed by charting biological rhythms.[51]

Chrononobiology may be used as an aid in treating patients. Biological rhythm charts can monitor the effectiveness of various treatments. For example, evidence shows that antidepressant drugs work best for people with slower biological clocks.[52] Chronobiology may even be able to save lives by determining the time of day when chemotherapy for cancer will be best received by the body.

While chronobiology is still in the formative stages, there are no pseudoscience indicators warning us away as they did with biorythms. Occasionally people get carried away and make suspicious-sounding statements about biological rhythms influencing "everything from basic physical functions to feelings about a particular shade of lipstick."[53] But on the whole, chronobiology appears to be an area of study that is being examined in a way that meets the criteria of legitimate science.

FIVE

BATTLEGROUND: PLANET EARTH

In the early years of the twentieth century, French scientist René Blondlot intrigued the science world with his discovery of N rays.[1] These mysterious rays could be detected by the way they intensified an electric spark through which they were beamed.

Blondlot's discovery was quickly confirmed by more than twenty French professional scientists. In 1904 French scientific journals published fifty-four papers dealing with N rays, compared to only three on another relatively recent discovery, X rays.

Some scientists, however, had trouble detecting N rays, and many of them visited Blondlot's laboratory to learn more about his work. Among the curious was American scientist R.W. Wood. One of Blondlot's experiments used a prism to bend N rays onto a phosphorus screen. Although Blondlot insisted that the N rays were showing up on the screen, Wood was unable to see anything. Blondlot's explanation was that special skill was required to know exactly what you were looking for.

Unknown to Blondlot, Wood removed the prism. When Blondlot continued to describe the effect of the N rays on the phosphorus screen, Wood knew something was wrong. Within a few years the evidence was overwhelming that N rays did not exist.

Had Blondlot simply admitted his mistake, N rays would have been just another example of a scientific error. But Blondlot compounded his error by resorting to pseudoscientific techniques. He explained away the lack of confirming evidence by insisting that the detection of N rays was an art that only he and a few others had mastered. Rather than objectively examining the evidence, he made his proposals increasingly complex to explain away the evidence.

Blondlot had slipped into the practice of pseudoscience, and he argued his cause persuasively for a while. He could not get away with it, though, because he was fighting on a battleground that favors science over pseudoscience—the surrounding world. Unlike the enormously complex and variable human mind, most of what goes on in the world around us can be broken down into elements simple enough to study. The environment in which we live is accessible and can be closely studied and tested. In an area where facts are accessible and predictions can be easily tested, pseudoscience has trouble getting a foothold.

ALCHEMY

The study of our surrounding world has been riddled with mistakes throughout the ages. For example, in prescientific times, Aristotle taught that all matter was made up of four elements: earth, air, water, and fire.[2]

The craft of alchemy grew out of this nonscientific world view. If all things were made of four elements, the only difference between substances would be the proportion of those elements. If that were true, then it should be possible to change one substance into another simply by altering the proportions. It could be demonstrated that water could be turned into "air"

A woodcut of an alchemist in his study.

simply by heating it. Why couldn't some base metal be turned to gold by some similar chemical process? Many alchemists tried, and over the course of several hundred years many claimed to have succeeded.[3]

The observations of alchemists provided a foundation for the science of chemistry. But alchemy does not exist as a science today because its assumptions were too easily disproved in the nineteenth century. Overwhelming evidence showed that there were many more elements than just four, and the science of chemistry pointed up the overwhelming nuclear forces that prevented base metal atoms from changing into gold atoms. Alchemy could not survive as a pseudoscience because it operated in areas where overwhelming evidence is too easily obtained.

COLD FUSION

A more recent example of scientific error is the cold fusion controversy.[4] On March 23, 1989, chemists Stanley Pons and Martin Fleischmann of the University of Utah announced that they had produced nuclear fusion in a test tube at room temperature. If this were true, it would rank as one of the most important scientific discoveries of all time.

Nuclear fusion involves forcing together the nuclei of atoms with such pressure that they fuse. The mass of these fused nuclei is less than that of the nuclei when separate. The lost mass is converted into energy.

According to calculations first worked out by Albert Einstein, a tiny amount of mass converts into an incredible amount of energy. If fusion could be accomplished on a large scale, it could provide a virtually unlimited supply of energy without creating the poisonous radioactive byproducts of today's nuclear fission energy plants.

Scientists Stanley Pons (left) and Martin Fleischmann display a sample of the flask in which they claimed to have created sustained nuclear fusion reactions at room temperatures. Their work was later discredited.

Unfortunately, a tremendous amount of energy is required to overcome the natural repulsive forces that keep atomic nuclei apart. Up to the time of the Pons and Fleischmann report, a tiny amount of fusion had been achieved only at temperatures of more than 100 million degrees Celsius. Far more energy was required to achieve fusion than was produced by it.

Pons and Fleischmann, however, claimed to have produced fusion at room temperature on an ordinary laboratory table, which would make the process commercially feasible. They started off on the right scientific track, presenting their findings and opening them up for peer review (criterion #5). They fully expected others to be able to duplicate their results.

But before the process got too far, cold fusion began to fail some of the basic rules of science. In violation of the scientific criterion that requires well-defined claims (criterion #2), the cold fusion chemists were vague about exactly how they conducted their experiments. A few scientists were able to duplicate their experiments, but most were not. Ten researchers at the Harwell Laboratory in Great Britain spent three months and half a million dollars trying to repeat the cold fusion success and failed.[5] The cold fusion miracle failed the repeatability requirement (criterion #4).

Those whose results did not support the chemists were told that they were "not doing it right." Yet chemists were still not given any more detailed descriptions of what Pons and Fleischmann actually did in their experiments.

A major tipoff that the effort was falling out of the domain of legitimate science came with the evidence that some of the data was being adjusted to suit the theory, a violation of criterion #8. Physicists pointed out that the fusion experiments should have produced gamma rays with an energy of 2,224 KeV (or 2,224,000 electron volts). Yet the initial reports indicated that the gamma rays were measured at 2,500 KeV. After physicists pointed out that this was a major discrepancy, a published paper on the cold fusion experiments suspiciously reported the gamma ray measurement as 2,224 KeV.[6] One physicist was threatened with a lawsuit for reporting experimental evidence against cold fusion.

Within two years, the report of cold fusion was

largely discredited. A U.S. Department of Energy committee concluded that no persuasive evidence existed to support the claim that cold fusion had been achieved.[7] The reported success appeared to be an illusion produced by a number of errors.

As with N rays, a scientific mistake had shown signs of degenerating into the realm of false science (what one writer referred to as "wishful science"[8]), only to be pulled up short by the massive weight of evidence. In the world around us, where the area of study is limited in scope and presents few variables, science can usually overwhelm pseudoscience before it gets established. In such a setting responsible scientists, even those who stand to lose prestige, waste little time in acknowledging their own errors and in squelching the claims of pseudoscience. But not always. Pseudoscience crops up wherever doubts and mystery exist.

EARTHQUAKE PREDICTION

On December 3, 1990, many residents of southeastern Missouri awoke with a gnawing sense of fear. Factories and schools were closed as a precaution against the possibility that a violent earthquake would strike on that day.

The uproar was caused by Dr. Iben Browning, a scientist, inventor, and business consultant from New Mexico, who claimed to have the ability to predict earthquakes. According to his tidal force theory, the gravitational pull of the sun and moon could, under the right conditions, create a subtle bulge in the Earth. This could trigger an earthquake on any fault line where pressure was building toward the rupture point. Browning noted that tidal conditions would be right on December 3 for triggering an earthquake. He identified the New Madrid fault line, which runs through southeastern Missouri and northeastern Arkansas, as the place where that earthquake would occur.[9]

Geological experts insisted that the New Madrid prediction had no scientific basis and should be ignored. But Browning had a track record for predicting earthquakes that could not be so easily dismissed. The *New York Times* reported that Browning had predicted, a week in advance, the 1989 earthquake that shook the San Francisco Bay area. Another paper gave Browning credit for predicting the time of that quake to within six hours. With that kind of evidence supporting Browning, many felt they could not take the risk of ignoring his warning.

December 3 came and went peacefully. No earthquake. Browning was wrong and, belatedly, everyone knew it.

How could the public possibly have known whether his theories had any validity? Is there any way for people to avoid getting taken in by these kinds of predictions?

In order to answer those questions, we first need to determine whether Browning simply made a scientific mistake or whether he was engaged in pseudoscience.

On the surface, Browning presented himself as a believable scientist. Well-educated, intelligent, creative, he presented a logical explanation for his predictions (criterion #1). His claims were backed up by the director of the Southwest Missouri State University Information Center. His basic information about the vulnerability of the New Madrid fault was correct. In 1811 and 1812, the area that he targeted had been jolted by three of the largest earthquakes in U.S. history. Earthquake specialists acknowledged that this location was overdue for some sort of disturbance. Further, more than one seismologist had speculated over the years as to whether tidal pull could have any connection with earthquakes. Browning knew his subject well enough to accurately forecast that the tidal forces would be at their peak on December 3. According to proper scientific procedure he had made a well-defined claim of what would happen on that date (criterion #2).

Yet telltale signs of pseudoscience were in plain sight. Browning's most glaring violation was a lack of peer review (criterion #5). The self-taught climatologist developed his methods and theories on his own. His work was not submitted to other professionals to be reviewed or tested. In other words, Browning provided no reason for believing him other than that he said he was right.

Peer review would have revealed that Browning's success in forecasting earthquakes was based on flimsy evidence. Six weeks before the predicted New Madrid quake date, the U.S. Geological Survey finally performed a review. After gathering evidence on their own, including transcripts and videotapes of his talks, they found that Browning's 1989 earthquake prediction had not even mentioned North America, let alone the San Francisco Bay. In violation of scientific criterion #2, he had made his reputation based on vague predictions and, in the tradition of pseudoscience, had accepted credit for more than he had actually predicted.[10]

Browning's main support for his theory rested on the fact that, although he did not say where it would occur, he *had* predicted a large earthquake on or about the date of the San Francisco quake. Scientific criterion #9 warns, though, that science does not accept coincidence as proof. An analysis of the frequency of earthquakes shows that this coincidence was hardly striking. In fact, predicting earthquakes is a favorite ploy of psychics because, since earthquakes occur somewhere in the world an average of every three days, they stand a good chance of being right. According to one earthquake expert, "We constantly deal with seers, channelers, experimenters, and backyard seismologists who are making predictions."[11] After examining Browning's five-year record of predictions, the

U.S. Geological Survey concluded that his method was no more accurate than sheer guesswork.

By the time this news came out, however, the damage had been done. The press had assisted Browning in violating the standard of caution in evaluating evidence (criterion #7) by writing headline stories trumpeting his untested theories based on the skimpiest of evidence. Confusion and misinformation had spread, not to be cleared up until December 3 pronounced an unmistakable verdict on Browning's prediction. But the signs of pseudoscience had been there all along.

CREATION SCIENCE

A recent survey of Ohio high school biology teachers reported that 22 percent of those responding said that they taught "creation science."[12] Creation science uses scientific techniques to support the view that life on Earth developed according to a particular interpretation of Genesis, the first book of the Bible. While this view receives support from a large segment of the U.S. population, it is rejected by most biologists.

Is this strictly a disagreement between two different groups of scientists over how to interpret the data? Or are we dealing with pseudoscience?

The history of the controversy provides some strong clues in this sensitive issue. It all began in 1858, when Charles Darwin concluded from the results of his study that species evolve from more primitive species over long periods of time.

Obviously no one can go back in time to observe what actually happened. But since Darwin's time, biologists and geologists have been testing the principles and assumptions on which evolutionary theory is based, including:

1. Natural selection—the tendency of certain traits

to be passed on from one generation to another because they are best suited for survival in a particular environment.

2. Adaptive radiation—the tendency of species to develop a wide variety of physical traits in adapting to different environments.
3. The tendency of isolated habitats to produce more different species than those areas where individuals of a species freely interbreed.
4. The accuracy of dating techniques that determine the age of an object by measuring the deterioration rate of certain molecules.
5. The pattern of simpler animal forms occurring earlier in the fossil record than more complex forms.

While scientists were testing these principles and finding a great deal of evidence to support them, heated opposition arose from outside science, primarily from those who felt that the theory of evolution conflicted with their religious beliefs. In 1925 the state of Tennessee passed a law forbidding any state-funded school to teach any "theory that denies the story of the Divine creation of man as taught in the Bible." Other states followed suit.

In 1968 the U.S. Supreme Court unanimously ruled such laws unconstitutional because they violated the First Amendment ban on state establishment of religion. Evolution opponents in Arkansas then passed a law in 1981 that required biology teachers at public schools to give equal time to "creationism" as an alternative to evolution. Again, the federal courts struck down the law as a form of government-established religion.

At about the same time, Louisiana creationists took a new line of argument. In passing a law similar to that of Arkansas, they argued that their stance on the origins of species was not a religious belief but was a science. They presented scientific evidence to support their view. Their expressed intent was merely to require a balanced presentation of "evolution science" and "creation science." In the case of *Edwards* v. *Aguillard* in 1987, the Supreme Court rejected this law as well.[13]

This brief history highlights a glaring clue that links creation science with pseudoscience. However closely it may follow certain rules of science, creation science strays far from several scientific criteria. The most obvious is that it arose from a nonscientific, nonverifiable belief system (criterion #3) and remains committed to that belief system regardless of evidence, in direct contradiction of scientific criterion #8.

Nonscientific belief systems can be good and they can be valuable, but they do not mix well with science. Strongly held beliefs are not easily persuaded by facts. Professionals in all areas of science must guard against getting too wedded to their own pet theories, let alone beliefs, to be objective. Strong beliefs make it desperately important for the holders of those beliefs to prove they are right. With so much at stake, an impartial scientific investigation is virtually impossible; investigations focus on trying to prove a particular point rather than to seek the facts.

Furthermore, the evidence collected by creation scientists has not passed the test of peer reviewal (criterion #5). The great majority of biologists reject the claims on the basis of evidence. Geological dating techniques, for instance, show the earth to be much older than many creation scientists claim.

Someday shocking new evidence may reveal that evolutionary theory is all wrong. If that happens, biologists will have to change their minds. That is how science works. But because most biologists believe there is a convincing case for evolution, creation science advocates have repeatedly passed laws to make people accept their arguments rather than relying on the force of evidence. This further contradiction of the basic criterion of scientific objectivity (criterion #8) places creation science squarely outside the boundaries of true science.

ECONOMICS

While phony science is relatively easy to isolate and identify when it deals with the measurable physical world, detection becomes more difficult when human interaction is thrown into the equation. The social sciences, which study how people interact with their environment and with each other, deal with erratic data. This is owing largely to the nature of intelligence.

The actions of animals that rely on instinct rather than intelligence to govern their actions are relatively easy to predict. Instinct is a built-in, programmed response to certain situations. Intelligence, however, gives a creature the flexibility to respond to a limitless number of situations and variables. The responses are not as predictable as instinct because they are based on an interpretation of an incredibly complex mass of information gathered from the senses.

Economics is a social science that studies how people interact with the world to create and distribute products and services. If those interactions were based on instinct, they would be predictable, and economics would be a reasonably simple science. But because those interactions are governed by intelligence, they can be quite unpredictable. Unpredictability means doubt and mystery—the very elements on which pseudoscience thrives.

So what are we to make of economic projections? Is it possible to make accurate predictions about the future of the economy that meet the rigid requirements of science?

Obviously, many economists make such predictions. Business sections of newspapers devote a great deal of space to the subject of what will happen in the future. Various experts offer their opinions on what the economy will do in the coming year: what level of inflation to expect, what the gross national product will be, whether employment rates will remain stable, whether interest rates will rise or fall, in which direction the stock market is headed. Many corporations hire economic experts to help steer their businesses through the murky waters of the future.

Economists use scientific terms, cite reams of statistics, and bring out impressive charts to back up their claims. Some go as far as to describe rules and mechanisms that govern the economy. In the 1970s, for example, Arthur Laffer gained a widespread following for showing that once you reach a certain level of taxation, any higher taxes will actually result in less revenue.[14]

Suddenly, though, economic gurus have fallen out of favor. Companies have recently been firing many of their economic experts. In 1990 the Continental Bank of Chicago fired eleven of their twelve forecasting sages. The reason for the firings was evidence showing that economists were not very good at making certain kinds of predictions. Stephen McNees, vice president of the Federal Reserve Bank of Boston, kept his own scorecard to match economists' predictions of inflation and gross national product growth against the actual result. He discovered that the experts missed virtually every recent shift in the economy.

McNees also studied the predictions of twelve well-known forecasters over the course of a decade to see which methods were most accurate. His analysis showed that none of them was consistently better than any other.[15] Other economists have made careful investigations of the ways that the stock market works and have determined that there is no reliable way of predicting what the market will do.[16]

Nobel Prize–winning economist Kenneth Arrow was not surprised. He explained that "when you have a complex open system, and ours has a great many interrelations, predictions become exceedingly difficult."[17] Economists arrive at their predictions by accumulating data from the past, measuring current conditions, and then comparing current conditions with the accumulated data. But there is no guarantee that the future will follow the pattern of the past. Too much information on all sorts of unpredictable people is needed. An economist can assemble mountains of data in the world and have it analyzed by computer to determine where the world economy is headed, only to have it all go up in smoke just because Saddam Hussein gets a sudden urge to invade Kuwait.

Is economics a science? If so, is it still a new-frontier science where all kinds of honest mistakes are likely to be made until further research reveals its secrets? If not, what is it? Is it pseudoscience?

In many cases, the study of economics follows the rules of science. Economists have demonstrated that they can perform useful services when their field is narrowed down to a manageable scope. Economists can be valuable in assessing, for example, the future prospects of an individual industry or company.

Science criterion #7 tells us that scientists must be careful in examining evidence and slow to make unsupported claims. Many economists show that sort of caution. In fact, acknowledging the complex, unpredictable aspect of their profession, many economists hesitate to call their subject a science. Pseudoscience is anything that pretends to be science but is not. If economists do not claim to be doing science, by definition, they are not pseudoscientists.

But those who do not exercise that sort of caution, and who claim a scientific basis for their predictions when there is none, run the danger of misleading their clients and the public. One business analyst summed up the rise and fall of corporate economic forecasters by saying that economics "oversold its ability to peer into the future."[18] That's more than a simple mistake in calculation—that's pseudoscience.

SIX

BATTLEGROUND: THE COSMOS

While sailing to England in 1930 to pursue his studies in astronomy, Subrahmanyan Chandrasekhar passed the time by working out brain-teasing mathematical exercises concerning collapsing stars.

Astronomers of his time had determined that when a star burns out, the force of gravity produced by its tremendous mass causes it to collapse into a small, incredibly dense ball. In checking the calculations on which that theory was based, the nineteen-year-old scholar from India reached a stunning conclusion. If Chandrasekhar's figures were correct, the tremendous gravitational force of a very massive star could even overcome the outward electromagnetic force that keeps atoms intact.

The implications of this were mind-boggling. If there were no force strong enough to stop the contraction, what would prevent the star from contracting indefinitely? Since the contracting star did not lose mass, what would prevent it from retaining the crushing gravitational pull no matter how small it became? Chandrasekhar's calculations pointed toward the probable existence of stars that were infinitely dense. These stars took up virtually no volume yet had gravitational forces hundreds of times greater than the sun, so strong that even light could not escape.

Chandrasekhar spent two years sharing his findings with the best astronomers in the world. In January 1931 he was invited to present his research at a meeting of the prestigious Royal Astronomical Society of England. Although aware of the astounding possibilities his calculations had opened up, Chandrasekhar did not try to impress people with exotic projections of what his discovery meant to the world. He simply presented his evidence. As for what happened to the collapsing star, he said simply that "one is left speculating on the possibilities."

When Chandrasekhar had finished, Sir Arthur Eddington strode to the podium with all the self-assurance of a man firmly established as the greatest English astronomer of his time. To Chandrasekhar's horror, Eddington launched into a speech ridiculing the young student's work. There was no such thing as the degenerating atoms that Chandrasekhar had proposed, he said. The whole idea was nonsense! "I think there should be a law of nature to prevent a star from behaving in this absurd way." Eddington explained away the findings by implying that Chandrasekhar had made mistakes in calculation.

Chandrasekhar tried in vain to salvage his reputation. But, stifled by the power of Eddington's reputation, he had to give up and he moved on into other areas of astronomical research.[1]

Because of Eddington's enormous influence, the realm of black holes, which was on the verge of discovery, remained shrouded in confusion for several decades. Not until the 1960s did most astronomers accept their existence. Since then, more data has been uncovered to verify black holes. Astronomers have repeatedly observed extremely powerful gravitational effects on stars in areas where no visible stars are present.

ASTROLOGY

The case of Chandrasekhar and black holes illustrates the problem with coming to grips with the reality of outer space. Even a respected astronomer such as Eddington found that his concept of logic failed him. Outer space deals with numbers and distances far outside of human experience, with concepts such as infinity that are beyond our comprehension. In a realm that is so inaccessible and beyond our imagination, what point is there in depending on logic for guidance?

A growing number of people have abandoned logic as a guide for understanding what is real in the great beyond. The forms of astrology with which the general public is familiar have little to do with logic and rationality. Astrology is the study of how the stars influence people and events on the Earth. Its most popular use today is in the construction of horoscopes that advise people how to conduct their lives.

The idea that a person's fate is determined by the position of the planets is as logically bewildering as the theory of black holes is to most of us. We see no obvious rational evidence that leads to the inevitable conclusion that a person's character should be shaped by the position of the sun in the sky. Could this influence be caused by gravity? Hardly. As one scientist points out, the "obstetrician hovering near the infant exerts a much greater gravitational pull than the nearest planet."[2]

Yet recent surveys have shown that more than a quarter of the United States population and upwards of one billion people in the world accept this kind of astrology as valid.[3] The percentage of young Americans who believe in astrology is even greater, about 50 percent.[4] An estimated 1,200 newspapers in North America print personal horoscopes for their readers. In fact, astrology has far more popular appeal than its

A meteor falls during the 1966 Leonid meteor shower. In ancient times, meteors were often believed to be omens.

cousin, astronomy. According to science author Carl Sagan, the United States has approximately 1,500 professional astronomers whose long-distance observation and measurements, mathematical calculations, and space probes try to determine the order and structure of the regions beyond the Earth. Their numbers are dwarfed by the 15,000 professional astrologers whose eyes are cast toward the heavens for a different purpose.[5]

But are the principles of astrology any more strange than some of the findings of astronomy? After all, where did Eddington's logic get him? This leaves the

A zodiac chart

disturbing question of how do we detect pseudoscience if we can't even use the first basic scientific criterion of logic?

The mistake here is in confusing logic with comprehension. Chandrasekhar's theory pointed to some possibilities that were extremely difficult to comprehend. That is a far cry from saying that the possibilities were not logical or rational. In fact, Chandrasekhar's theory depended entirely on rational thinking every step of the way. It was achieved by applying accepted principles of physics and mathematics, including those developed by Albert Einstein, to information on stars obtained by Eddington himself. No matter how strange the result may seem in astronomy, it follows from logical, rational reasoning.

Can the same be said for astrology? Do the conclusions of astrology follow from logical, rational reasoning? In order to answer that we need to look first at the origins of astrology. Historians have traced astrology back as far as the eighteenth century B.C.[6] Ancient sky watchers in what is now called the Middle East were intrigued by a small number of exceptional stars that wandered through the sky.

In the country of Babylon, a society saturated with superstition, the sky watchers puzzled over what these wanderers (known today as planets, from the Greek word for "wanderer") could be. Like most mysterious and unexplained elements of the Babylonian world, the wanderers were connected with the supernatural. The explanation that arose was that the planets were either gods or messengers sent by the gods.

With the development of a calendar, the Babylonians were able to keep charts of when and where the wandering stars appeared in the sky. Their precise records offer one of the first known examples of systematic scientific data-gathering. From their observations, stargazers began to notice that these planets moved in predictable patterns.

The Babylonians, particularly the royalty, were constantly looking for ways to foretell the future to take the guesswork out of difficult decision making. For example, many of them believed the soul was located in the liver, and this led them to base countless predictions on their analysis of sheep livers.[7] Others looked to the stars. Astrologers began to make lists of lucky and unlucky days and studied the sky to see how the planets were aligned on those days. This study led them to draw conclusions about how good and bad fortune corresponded to the alignment of the stars. From this they developed the art of reading the future from the movements of the planets.

Personal horoscopes entered the scene between 400 and 500 B.C. They were actually an adaptation made by the Greeks, who admired the mathematical beauty and precision of the Near Eastern astrological charts.[8] Horoscopes were particularly popular in places where planets were considered gods. In fact the names we use today for planets are the names of the gods these planets represented. Mercury, for example, was the fleet-footed messenger god, and was associated with the planet that moved most swiftly. The planet with a reddish hue reminded the Romans of Mars, the fierce god of war.

Modern astrologers claim that the deep-rooted, enduring history of the science of astrology is proof of its validity, but the nature of its origins hardly inspires confidence. Although astrology relied on scientific observations and mathematics to chart the positions of the heavenly bodies, it developed as a system of magic, was based on a belief system of superstition, and came to be closely connected with mythology—not a promising résumé for an enterprise that seeks acceptance as a rational, logical science in the twentieth century.

Yet modern horoscopes continue to flourish. The most common types are based on the zodiac, which was developed by ancient Babylonians, Assyrians, and Egyptians as a means of measuring time.[9] The creators of the zodiac assumed that the sun traveled around the Earth. They divided the circle that the sun traveled during a year into twelve sections. Each section was named after the dominant constellation or group of stars in that section of the sky—Aries the Ram, for example.

Alfred Hug claims that astrology affects the stock market. Note the similarities in his computer-generated graphs.

A person's individual horoscope is centered on the section of the zodiac the sun was in when the person was born. An Aries is simply a person who was born between March 21 and April 19, when the sun was in the Aries segment.

The premise of the horoscope is that the position of the sun in the zodiac determines an individual's personality and characteristics. Further, events and trends in that person's future can be predicted by skilled astrologers who understand the significance of various patterns and positions of heavenly bodies in the night sky.

Many people swear to the accuracy of their horoscopes. Others have described astrology as "a massive fraud that costs Americans billions each year."[10] Who is right?

Measured against the scientific criteria, horoscopes fall into trouble immediately with the standard of drawing logical conclusions that inevitably follow from the data (criterion #1). Beyond their suspect history and the fact that the zodiac was based on the mistaken notion that the sun revolved around the Earth, there is no obvious reason why planetary movements should influence human character.

Further, the horoscope premise contains some glaring flaws in logic. For example, as viewed from the Earth, the position of the sun relative to the constellations does not stay constant. This means that on any given date, the sun is no longer in the same segment of the zodiac that it was on that date a couple of centuries ago. So a person born between March 21 and April 19, whose horoscope is given the characteristics of an Aries, may actually be a Pisces based on the position of the sun when the original charts were drawn up.[11] If modern horoscopes are based on the ancient findings, then logic tells us that either today's horoscopes or the originals were wrong.

Astrologers counter this by saying that true astrology does not involve the sun constellations but focuses primarily on the planets, whose position relative to the Earth does not change. But even in this case a logical flaw remains, one that was pointed out centuries ago. If fate and personality are dictated by the position of planets, then all twins should have identical fates and personalities. Identical twins, however, seldom have identical fates, and fraternal twins frequently show vastly different personalities.[12]

Some astrologers respond to this by declaring that "the stars incline; they do not compel," that birth charts indicate only a person's potential.[13] In other words, the heavenly bodies exert an influence but that influence does not always show up because people have a limited ability to control their own destiny. This kind of argument immediately triggers pseudoscientific warnings because it violates the scientific standard of falsifiable hypotheses (criterion #3). The statement cannot be disproved because it adjusts to fit all data, even contradictory data. No matter what sort of test you devise, no matter what results you obtain, no matter what observations you make, that hypothesis can explain it.

Astrologers get around a number of problems by producing increasingly complex explanations that account for the negative result. According to the Code of Ethics of the American Federation of Astrologists, an "opinion cannot be honestly rendered unless based on a horoscope cast for the year, month, day, and time of day plus geographical location of the place of birth."[14] But accurate time telling is a fairly recent invention. How were horoscopes determined before the invention of clocks and watches to tell them exact birth times? Logically, the art of reading the future in the stars must be based on centuries' worth of data and conclusions gathered by people who could not "honestly render" an opinion.

When someone gives increasingly complex hypotheses in the face of negative results, it is a strong indication that the explanation is being cobbled together to fit whatever results occur. This violates the scientific criteria of objectivity (criterion #8), which is difficult to maintain when strongly held beliefs are challenged by fact.

Leaving all this aside for the moment, do horoscopes somehow produce results? Are the predictions accurate? Where is the evidence?

The nonfalsifiable aspect of astrology makes evaluation difficult. But an even greater problem is that horoscopes often violate the scientific criterion concerning well-defined claims (criterion #2). This is a necessary characteristic of pseudosciences that deal with the future. Specific predictions are exposed to the light of investigation. The clearer and more specific the prediction, the easier it is to prove or disprove. Pseudoscientific forecasters leave themselves a way out in case something goes wrong: They make their predictions so vague or ambiguous that they cannot be proved false.

A classic example of this, which, curiously, is used by fortune-tellers as evidence of the validity of their craft, occurred in the sixth century. A Greek king named Croesus, trying to get up the nerve to attack his enemies in Persia, asked an astrologer if an attack was wise. Croesus was advised that if he crossed the river into Persia a great empire would fall. Croesus took that as supernatural support for his venture and proceeded. He and his armies were promptly destroyed by the Persians.

Did this discredit the seer who gave such terrible advice? No, the fortune-teller left himself an out. He didn't say *which* empire would fall. An empire fell, although Croesus was not expecting it would be his own.[15] Worded carefully, predictions can be made to fit a great number of results.

Study any newspaper horoscope and examine the nature of the predictions. In the words of one critical publication, such predictions are generally either "self-fulfilling, good bets, non-specific, or right either way."[16] As an example, a horoscope that says you must exercise caution in business dealings this week is giving sound counsel that applies to anyone at any time. A warning that "insensitivity on your part will lead to troubles in romance" is on similarly safe ground. If you have a lover's quarrel, the prediction was right on the money. If you do not, you can thank your horoscope for preventing it.

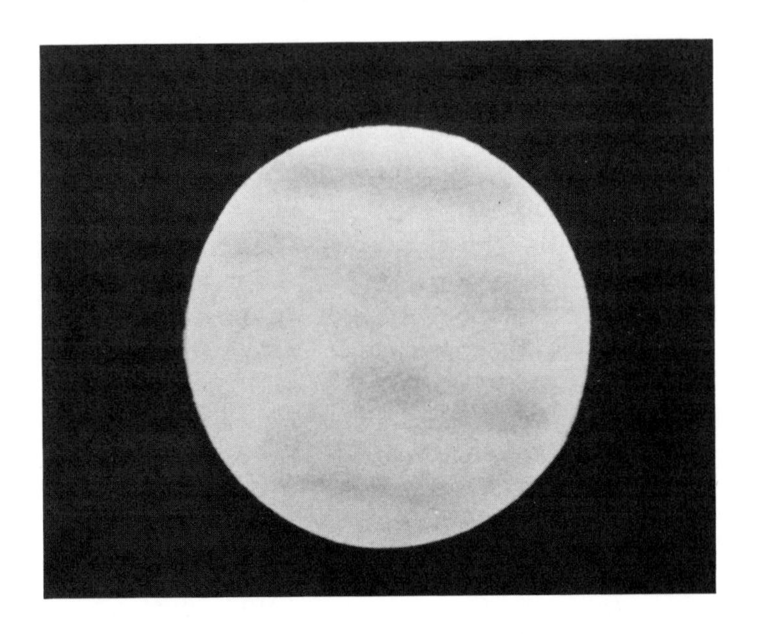

In the nineteenth century, irregularities observed in the orbit of Uranus led to the discovery of Neptune (shown here). This is an example of how astronomical *predictions differ from* astrological *ones.*

Contrast this with the predictions of an astronomer engaged in legitimate scientific research. In the early nineteenth century, astronomers were puzzled by irregularities in the orbit of the planet Uranus. Several proposed that these orbital quirks were caused by something, probably an unseen planet, that was exerting a gravitational pull on Uranus. In the 1840s Englishman John Adams and Frenchman Urbain Le Verrier made careful calculations, based on the orbit of Uranus and the known effects of gravity, to predict where this unseen planet must be located. Astronomers then looked in the sky where Adams and Le Verrier had steered them and immediately discovered the planet Neptune in almost precisely the spot where it had been predicted.[17] Astronomers were able to make such a bold, specific prediction because they had carefully collected enough evidence to back it up.

What of those astrologers who do make specific predictions? And what of the many people who have followed their horoscopes for years and believe they have seen results? Don't they offer proof that astrology works?

Not according to the rules of science. Science cannot accept anecdotal evidence as convincing proof (criterion #10). The stories used to verify the success of a prediction cannot be examined in their entirety to see if they can be confirmed or discounted. Too many variables exist: does the number of correct predictions fall within the range of random chance; were the predictions self-fulfilling, or vague enough to be interpreted several ways? The human mind, as discussed in chapter 3, is highly susceptible to the power of suggestion. Tests have shown that those who know the identifying characteristics predicted by their astrological sign are more likely to claim those characteristics than those who do not.[18]

Independent, objective tests tell a different story than that given by anecdotal evidence. Astronomers Roger Culver and Philip Ianna evaluated more than 3,000 specific predictions made by well-known astrologers. They concluded that 90 percent of the predictions failed to come true.[19] Bernard Silverman, a psychologist at Michigan State University, studied the records of nearly 500 couples who divorced in Michigan in 1967 and 1968 and found no correlation with the compatibility predictions of astrologers.

Seven independent studies performed between 1978 and 1986 asked subjects to choose which of two or more astrological birth charts fit them the best. According to the claims of astrology many of the subjects should have been able to do so. All seven studies found that subjects' ability to match their own personalities with the correct birth chart was no better than chance.[20]

In another test, half of the subjects were given birth charts compiled for their date of birth and asked to rate how accurately the charts fit them. The other half of the test group were given birth charts as nearly opposite as possible to their correct birth charts and asked to rate how accurately the charts fit them. Under these blind conditions, the results were virtually identical. The opposite birth charts were judged to be no less accurate than the real charts.[21]

Perhaps the most compelling evidence against the accuracy of astrology was gathered by astrophysicist Shawn Carlson. Carlson sought out thirty American and European astrologers who were rated by their peers as among the best in the profession. He asked them to study the birth charts of 116 people. The astrologers were then given three personality profiles for each of the 116 clients. One of the profiles was accurate, the other two were selected at random. Without seeing or knowing anything about these people, the astrologists were asked to choose the profile that best matched the birth chart.

Carlson worked closely with the astrologers to set up a test that was as fair as possible. He encouraged their advice and used every reasonable suggestion. His conclusion from the test: the astrological predictions simply were not accurate.[22]

Perhaps the one astrological result that is hardest to explain away is a study undertaken by Michel Gauquelin in 1955.[23] Gauquelin was a skeptic who wanted to determine for himself whether the claims of astrology were true. In making his study, he was determined to use scientific methods.

While his initial findings failed to turn up anything conclusive, Gauquelin was surprised to find an influence of the planets that stood up to his challenge. He matched the position of the planets relative to the horizon with the birth times of a number of prominent individuals. He found that certain planets favored certain professions. Scientists, for instance, tended to be born when Saturn was dominant, and sports heroes and military leaders were favored by Mars.

Gauquelin completed a study in 1960 that confirmed these findings beyond any statistical coincidence. Based on more than 20,000 cases, his study concluded that different planets were linked with different professions in a quirky way. The effect was minimal among "ordinary" people, but strong among the more famous or accomplished.

Baffled by the results, Gauquelin invited skeptical scientists to examine the evidence for themselves. Since 1960, three independent studies have been made on this subject. Two of the three found some support for Gauquelin's findings.

Does this mean that Gauquelin has scientifically proved a bizarre influence of the planets? Not yet. According to the scientific criterion #7, extraordinary claims demand extraordinary proof before they can be accepted. As no one has offered any explanation for how stars exercise this effect, the claim must be considered extraordinary. More evidence will be needed to establish it as fact. The scientific criterion of repeatability has not been firmly established. John McGervey, a physicist at Case Western Reserve University, researched the birth dates of the 16,634 scientists listed in *American Men of Science* and 6,475 politicians listed in *Who's Who in America.* He found the relationship between professions and birth dates to be totally random.[24]

Gauquelin's findings may well turn out to be a colossal error. Critics are particularly suspicious of a possible flaw in the studies: "famous" and "important" are not scientific concepts. This leaves researchers open to human error in selecting subjects with those qualities.

But at least the indicators do not show that Gauquelin is guilty of pseudoscience. He has followed the rules of science, including clear claims and peer review. Under those conditions it should be possible eventually to prove whether or not the effect he describes is real.

There are also conscientious scientists who have followed the rules in studying the effects of heavenly bodies on humans and on planet Earth. Some of them have been proven right. For example, scientists have provided ample evidence that the gravitational pull of the moon produces the tides. There are also intriguing reports that sunspots affect the ability of human blood to clot.[25] Such reports may be opening up a new frontier science of cosmobiology.

But these are exceptions. Astrology as commonly practiced today (including horoscopes) is choked with evidence of pseudoscience. The litter of clues should be enough to make people think twice before spending their fortune trying to find their fortune in the stars.

UNIDENTIFIED FLYING OBJECTS (UFOs)

Reports about strange flying spacecraft began surfacing shortly after World War II. The term "flying saucer" was first used by a private pilot flying near Cascade Mountain in Washington.[26] On June 24, 1947, the pilot reported seeing a strange flying object that looked "like a saucer skipping over water."

Nine days later something fell out of the sky and landed in a New Mexico desert.[27] Rumors circulated that it was the wreck of an alien spacecraft. The fact that the U.S. Army Air Force roped off the area five days later added fuel to speculation. Soon rumors began spreading that the bodies of four extraterrestrial beings had been found in the wreckage. According to the story, the government had sealed off the area to remove all evidence to avoid panic among the population.

Two British college students used trick photography to perform this hoax of a UFO flying over the Houses of Parliament.

A Zanesville, Ohio, barber and amateur astronomer Ralph Ditter took these photos of a purported UFO in 1966.

By the early 1950s, the extraterrestrial explorers apparently had grown more bold. People reported being kidnapped and held by aliens.

Since then, our world appears to have been regularly visited by alien explorers. If seeing is believing, then unidentified flying objects (UFOs) are a fact of life. More than one out of ten Americans has claimed to have witnessed a UFO.[28] Flying spaceships have been sighted all over the world by reasonable, intelligent people. Hundreds of people report that they have been abducted by alien beings and subjected to sometimes painful medical examinations before being returned home.[29]

Further, some claim there is evidence that our planet has been visited by alien astronauts in ancient times. Primitive cave paintings show what appear to be travelers hunched in spaceships. A stretch of ground near Lima, Peru, has been etched with more than 13,000 lines in more than 100 geometric shapes, which are said to indicate some sort of UFO landing field.[30]

How does the evidence of alien beings and flying spaceships hold up against the standards of scientific proof? The question is difficult to answer because of the wide range of claims surrounding UFOs. But the scientific standard of logical and rational thought (criterion #1) can at least be applied against the ridiculous hype of tabloid newspapers who claim Elvis Presley is an alien and those who claim UFO kidnappings in large cities are common.

Another level of common sense can be applied against *all* reports of UFOs. According to what we know about distances between stars, UFOs run into a simple problem of logistics. Light has been measured to travel at a speed of 186,000 miles per second. That works out to something like six trillion (6,000,000,000,000) miles per year. According to Carl Sagan, even if some kind of spaceship were constructed that could travel at that tremendous speed, it would still take 200 years to reach the earth from the nearest star in the galaxy.[31] Astronomers have also calculated that a ten-person spaceship traveling at 70 percent of the speed of light for a distance of five light-years would burn 500,000 times more energy than that consumed by the entire United States in a year.[32]

Logic says that such distances, and the time and energy required to travel them, present barriers that cannot be crossed even by a race of superior intelligence.

But suppose UFOs can traverse the time and distance in some way that is beyond our comprehension. Serious scientists have speculated on the possibilities. Is the evidence that is presented in support of their existence being obtained scientifically?

Much of it is not. On the extreme end, the claim that UFO abductions in urban areas are never witnessed because witnesses' memories are erased violates the scientific criterion about falsifiable hypotheses (criterion #3). What possible evidence could disprove that?

The more plausible UFO sightings occur randomly and last a short time, which means that scientists cannot return to verify or repeat the observation. This runs afoul of the standard of repeatability (criterion #4). The random nature of sightings does not necessarily indicate that UFO reports are untrue, but it does present problems in finding scientific support for it.

Support for the existence of UFOs relies heavily on anecdotal reports, which science cannot accept as conclusive proof (criterion #10). The mind and senses have ways of playing tricks on us, as can be illustrated by any number of optical illusions. The possibility for error in observation is shown by the fact that UFO investigators have found that Venus, the brightest of the planets, has been responsible for more UFO reports than any other known cause.[33]

UFO analysis has historically had a great deal of trouble with the scientific criterion concerning thorough and complete evidence (criterion #6). Of course, this could also indicate lack of research in a new frontier science. But the sensational appeal of UFO stories has helped produce the pseudoscientific trait of wild speculation based on extremely sketchy information. Odd land formations such as the etchings in Peru may defy a scientific explanation. But it is a huge jump to go on to suppose that just because their cause is unknown they have anything to with UFOs.

The characteristic caution of scientists in sifting evidence (criterion #7) also is absent in much of the discussion about UFOs. During the 1950s the U.S. Air Force took reports of UFOs seriously because of the nature of the cold war. Suspicious flying objects were checked out for fear that they might be Soviet spy satellites or other inventions. Air Force investigations, however, turned up nothing.[34] The suspicious object that crashed in New Mexico in 1947, for example, proved to be a weather balloon.

When reports of UFOs persisted and even multiplied, a group of scientists took the issue seriously enough to form the National Investigative Committee on Aerial Phenomena. Formed in 1954, this private research group examined claims about UFOs and concluded there was nothing to the reports.[35]

The Air Force reached the same conclusion, but in 1966 commissioned another group of independent scientists to double-check. This group, too, found no evidence for the existence of extraterrestrial visitors.

The many uses of fakery and trick photography to produce photographic evidence for UFOs indicate that some researchers are not interested in uncovering the truth, contrary to the criterion of objectivity (criterion #8). Honest UFO proponents further compound that error by disregarding evidence that rebuts their arguments. A group of UFO believers recently asked two investigators to check out some photographic evidence for UFOs presented by Ed Walters. The investigators, Rex and Carol Salisberry, had earned the respect and confidence of the group with a series of probing reports. The Salisberrys found that Walters was "adept at trick photography" and concluded that he had faked photographs. Instead of accepting the conclusion of an impartial investigator, the group who commissioned the report refused to accept it, and dismissed the Salisberrys.[36]

At least one organization has been conducting a proper scientific search for life outside the Earth. The National Aeronautics and Space Administration (NASA) has invested $100 million in the Search for Extraterrestrial Intelligence (SETI). The project employs more than a hundred physicists, astronomers, computer programmers, and technicians.[37] Using radio telescopes, including an enormously powerful one unveiled in 1992, SETI scans the sky twenty-four hours a day, listening for radio signals. Sometimes the search sweeps across the sky a half degree at a time; at other times it targets a particular area.

The scanners listen for radio signals or noises not commonly heard in nature. The reasoning is that if civilizations do exist out in the cosmos, there is a chance that they emit radio waves. Unlike the irregulular radio waves given off by stars, radio waves used for communication are sent in patterns. Any reception of patterned radio waves by radio telescopes would then indicate intelligent life.

Some scientists think that deliberate signals would blend in too well with background to be detectable. But SETI coordinator Thomas McDonough believes the idea is worth testing. Unlike most UFO investigations, which McDonough believes are done in a sloppy fashion by people who are poorly trained and easily fooled, the SETI experiment is based on a "testable prediction."[38]

Once a signal is located, it will be relocated again and again and examined carefully to make certain it is authentic.[39] This follows the scientific criterion of repeatability (criterion #4).

All this, according to McDonough, "is good science. If I'm right, SETI will probably succeed. If I'm wrong it will probably fail."[40]

Most astronomers refuse to rule out the possibility that there are other forms of intelligent life in the universe. Their reasoning is partially based on mathematical probabilities. Our galaxy contains several billion stars. The universe contains several billion galaxies. Out of that mind-boggling number of stars, what are the odds that only one star—our sun—has a planet with the required conditions for intelligent life? The tremendous distances and numbers of stars involved, though, make it impossible to test that speculation with current technology. The study of life in other star systems is a new frontier science about which scientists can say little. That is not to say that study of the existence of UFOs is pseudoscientific or nonscientific. The SETI project proves that the subject can be approached scientifically.

But if there are extraterrestrials piloting spaceships in our atmosphere, science has not been able to detect them. According to scientific criterion #7, extraordinary claims that aliens are visiting our world demand extraordinary evidence. All we have are a great number of random eyewitness reports of what people saw or thought they saw. That hardly counts as extraordinary evidence, especially in view of the fact that most of the UFO reports show all the earmarks of pseudoscience.

SEVEN

A PSEUDOSCIENCE QUIZ

Pseudoscience can take an almost endless variety of forms. Since no one knows what the next fad will be or what new evidence will be offered, it does little good to memorize the evidence in favor or support of everyone or everything that claims to be scientific. Rather, we must learn how to recognize the pseudosciences as they come, using the scientific criteria as a guide.

The following exercise offers a chance to test your skill at spotting pseudoscience. Six areas of research are briefly described. From the information given, using the scientific criteria (page 29) as a guide, try and determine in which of these categories each belongs:

A. science (correct)
B. science (mistaken)
C. science (new frontier)
D. pseudoscience
E. subject beyond the realm of science

1. During the 1950s several groups of American scientists probed the possibility that certain atoms could be stimulated into giving off light energy. They were look-

ing for some way to concentrate and focus this emitted light into a single stream.

In 1957 researcher Gordon Gould sketched out some rough ideas on how such a device could be built. Among his ideas was the use of an optical pump. This consisted of an excitable solid material surrounded by a coiled lamp. Light from the lamp would energize the atoms and cause them to release light of a single wavelength. This light would be reflected back and forth through the tube by mirrors on either end, and each time the atoms passed through the solid material they would produce further light emission. Eventually, a powerful beam of light would be created.

After calculating the incredible power that could be produced, Gould brainstormed a number of possible uses for this light-amplification machine, including heating and radar. Gould named his technique Light Amplification by Stimulated Emission of Radiation, or "laser" for short.

A number of other scientists recognized the possibilities of light amplification. From a wealth of research about the process that became available, Theodore Maiman constructed a working laser in 1960, using a synthetic ruby as the excitable solid material.

2. The Numerological Research Institute (NRI) reports the astounding success of a Dr. Forrest in tapping parapsychological forces. Dr. Forrest's powers are verified by eyewitness testimony from two dozen people who have benefitted from them.

The NRI is dedicated to making parapsychology work for the ordinary person and so they are presenting this special offer to you. From the extrasensory data he has collected, Dr. Forrest has calculated a list of lottery numbers that are guaranteed to win. For the nominal fee of $29.95, Dr. Forrest will provide you with a list of these numbers so that you can join the dozens who have already benefitted.

3. The "Big Bang" theory holds that the universe as we know it was created by a tremendous explosion of energy. Using the latest sophisticated equipment, astronomers have been able to measure more closely than ever the rate at which the universe is expanding.

The latest readings are very much in line with the Big Bang theory. Some people believe this proves that the creation of the universe was a random occurrence with no purpose. Others agree with one astronomer who was moved to comment that viewing the evidence was "like seeing God." Still others argue that the Big Bang theory is false because it does not account for the uniqueness of Earth and of human life among creation.

4. In recent years scientists have determined that the burning of fossil fuel has led to an increase in the amount of carbon dioxide in the atmosphere.

Many scientists believe this could have significant consequences for the world. Experiments have shown that increased levels of carbon dioxide in the atmosphere can block energy from the sun as it radiates off the earth, causing the "greenhouse effect."

Some warn that this greenhouse effect is producing global warming that will have a catastrophic impact on the world. This warming will melt the polar ice caps, raising the level of the oceans so much that they would drown all the coastal cities. Climates will change drastically in response to global warming and probably turn the Great Plains of the United States into a desert.

Others say that global warming is nonsense. No proof exists that carbon dioxide is raising the average temperature of the Earth. The Earth's temperature is not constant. Over the eons it has grown warmer and colder for reasons that have nothing to do with the greenhouse effect. This fluctuation of temperature over time makes trends difficult for beings as short-lived as humans to detect.

5. In the early 1960s, a Russian scientist discovered "polywater," an extremely rare molecular variation of regular water. Among its traits were a higher boiling point, lower freezing point, and greater stability than normal water.

The discovery attracted worldwide attention and sparked further research. At least seventy-five papers were published on the subject over the next decade. The presence of polywater posed a frightening possibility. If polywater came into contact with the natural water supply, the less stable water molecules would change into polywater. Since life on Earth is dependent on water, that could end life as we know it.

The initial research efforts, though, produced only tiny amounts of polywater. Skeptical scientists wondered if the results being reported in experiments were truly caused by polywater or simply by minute impurities. Suspicion grew when other scientists were unable to duplicate the original results. By the mid-1970s scientists had discounted the idea of polywater.[1]

6. Recently researchers have proposed the idea that many, if not all, of the body processes in humans are regulated by a pigment known as melanin. Melanin is what produces the dark color in skin. The ways in which melanin works are not completely understood. But according to some researchers, dark-skinned people are more advanced because their melanic system has continued to evolve while light-skinned people have lost that ability. The athletic superiority of African Americans in sports such as basketball provides one example of the benefits provided by melanin.

The influence of melanin explains white racism as a necessary defense developed by light-skinned people to protect them against a superior race.[2]

ANSWERS

1. A—The ideas were constructed logically from known principles. Specific claims were made that could be thoroughly tested. These claims were accurate enough that a great many useful devices have been built on the principles of light amplification for use in surgery, computing, communications, entertainment, and many other areas.

2. D—Although ESP claims are, by definition, not confined to logic, there is an element of logic (criterion #1) missing that is glaringly suspicious: If this group can select winning lottery numbers at will, why do they need your thirty bucks?

The secretive nature of Dr. Forrest runs contrary to the requirement of peer review (criterion #6). Further, the evidence is entirely anecdotal (criterion #10).

3. E—While evidence for and against the Big Bang may be scientifically gathered, such things as the purpose of life and who or what ultimately set in motion the series of events that led to creation of the universe cannot be addressed by science. These questions can be addressed only by religion and philosophy.

4. C—Claims are logical and clear, with evidence to support them, but are incomplete (criteria #6 and #7). At present, there is no way to obtain conclusive proof as to the amount of global warming that is taking place or its long-term effects. Further research and advances in measuring techniques may provide more evidence.

5. B—The only possible violation was of criterion #7, a clear signal that although a mistake was made, the rules of science were followed. Because the inquiry followed the scientific method, the mistake was easy to discover.

6. D—This presentation violates a number of scientific criteria. The idea that a pigment controls bodily functions does not logically follow from what science knows of the way the body works (criterion #1). A careful observer would have to ask what means have been provided for testing the theory (criterion #3). The conclusions presented require jumps in assumptions (criterion #6) and are supported by flimsy evidence, if any (criterion #7), particularly the link between melanin and the African-American domination of basketball. Further, this emotionally charged issue has potential for researchers to hold a personal bias toward the information (criterion #8).

EPILOGUE:

THE IMPORTANCE OF RECOGNIZING PSEUDOSCIENCE

"There is something fascinating about science," American humorist Mark Twain once observed. "One gets such wholesale return on conjecture out of such a trifling investment of fact."[1] Like so many others, Twain confused science and pseudoscience, but he described perfectly the problem with pseudoscience. Phony science allows conjecture to run well ahead of the facts.

Throughout history, horrible tragedies have resulted from conjecture running far ahead of the facts. Several centuries ago, many people found a connection between natural disasters such as hailstorms or the plague and the activities of witches.[2] Failure to challenge that conjecture and the flimsy evidence upon which it was based allowed more than 200,000 witches to be burned or hanged in Europe between the fourteenth and eighteenth centuries.

Failure to challenge the runaway conjecture of scientific-sounding "experts" has allowed some of the worst outbreaks of racism. A 1923 earthquake in Japan was blamed by some on Korean immigrants living in Tokyo.[3] The basic racial ideas that fueled Adolf Hitler's evil Nazi empire were fashioned in the laboratory of pseudoscience, built out of a few twisted, widely spaced facts and massive amounts of conjecture.[4]

Allowing conjecture to run well ahead of fact is as dangerous today as it has been in the past. Recently Nobel Prize–winning physicist William Shockley advanced his own supposedly scientific theory that whites are genetically superior to other races. The conjecture of pseudoscience repeatedly obstructs justice in the legal system by baffling jurors with impressive-sounding evidence from self-proclaimed experts. Every day, pseudoscientific conjecture bilks people of their money, robs them of their health, and even causes death.

Pseudoscience is not just harmless superstition. Learning to recognize it is not a trivial pursuit but a necessity if we are to make intelligent decisions for the future.

SOURCE NOTES

CHAPTER ONE:
False Information

1. Constance Holden, "Irrationality—Science Strikes Back," *Science,* April 13, 1990.
2. Ibid.
3. James Randi, "Help Stamp Out Absurd Beliefs," *Time,* April 13, 1992.
4. Holden, "Irrationality."
5. Peter W. Huber, *Galileo's Revenge: Junk Science in the Courtroom* (New York: Basic Books, 1991), p.4.

CHAPTER TWO:
Understanding Science and Pseudoscience

1. Tom Crouch, *The Bishop's Boys* (New York: W.W. Norton, 1989).
2. Terrence Hines, *Pseudoscience and the Paranormal* (Buffalo, N.Y.: Prometheus, 1988), p.15.
3. *Into the Unknown* (Pleasantville, N.Y.: Reader's Digest Association, 1981), p.53.
4. Thomas Hardy Leahey and Grace Evans Leahey, *Psychology's Occult Doubles* (Chicago: Nelson-Hall, 1983), p.5.
5. R.A. McConnell, *Introduction to Parapsychology in*

the Context of Science (Pittsburgh: University of Pittsburgh, 1983).

6. Taylor Stoehr, *Hawthorne's Mad Scientists* (Hamden, Conn.: Archon Books, 1978), pp.21, 22.

7. *The Skeptical Inquirer,* Fall 1991, p.54.

8. Kendrick Frazier, ed., *The Hundredth Monkey & Other Paradigms of the Paranormal* (Buffalo, N.Y.: Prometheus, 1991), p.26.

9. Holden, "Irrationality."

10. Hines, *Pseudoscience & the Paranormal.*

11. *The Skeptical Inquirer,* Summer 1991, p.382.

12. Ibid.

13. Paul Kurtz, ed., *A Skeptic's Handbook of Parapsychology* (Buffalo, N.Y.: Prometheus, 1985).

14. Hines, Pseudoscience and the Paranormal, p.4.

15. "Anti-Science Trends in the USSR," *Scientific American,* August 1991.

16. *Into the Unknown,* p.13.

17. "Anti-Science Trends in the USSR."

18. *The Skeptical Inquirer,* Winter 1992, pp.167–172; Hines, *Pseudoscience and the Paranormal.*

19. Aaseng, Nathan, *The Disease Fighters: The Nobel Prize in Medicine* (Minneapolis: Lerner Publications, 1985), p.15.

20. *The Skeptical Inquirer,* Winter 1992, pp.167–173.

CHAPTER THREE:
Battleground: Mind and Spirit

1. Michael Arvey, *ESP: Opposing Viewpoints* (San Diego: Greenhaven, 1989), p.48.

2. Hines, *Pseudoscience and the Paranormal,* pp.109, 118.

3. *The Skeptical Inquirer,* Winter 1992, p.173.

4. *The Skeptical Inquirer,* Fall, 1991, p.67.

5. James Alcock, *Science and Supernature* (Buffalo, N.Y.: Prometheus, 1990), p.10.

6. Arvey, *ESP: Opposing Viewpoints,* p.12.

7. Ibid., p.23.
8. Alcock, *Science and Supernature*, p.82.
9. Hines, *Pseudoscience and the Paranormal*, pp.82–86.
10. Arvey, *ESP: Opposing Viewpoints*, p.23.
11. Frazier, *Hundredth Monkey*, p.143.
12. Ibid., p.149.
13. Ibid., p.152.
14. Ibid., p.132.
15. *Into the Unknown*.
16. *The Skeptical Inquirer*, Fall 1991, p.80.
17. *Into the Unknown*.
18. James Randi, *The Mask of Nostradamus* (New York: Charles Scribner's Sons, 1990), pp.174–176.
19. Frazier, *Hundredth Monkey*, p.33.
20. Randi, *The Mask of Nostradamus*, pp.34–36.
21. Ibid.
22. Arvey, *ESP: Opposing Viewpoints*, p.54.
23. *Man, Mystic, and Magic*, vol. 24. (New York, Cavendish, 1970).
24. *The Skeptical Inquirer*, Fall 1991, p.35.
25. Ibid.
26. Ibid.
27. Alcock, *Science and Supernature*, p.1.
28. *The Skeptical Inquirer*, Fall 1991, p.52.
29. "Psychic vs. Skeptic," *Scientific American*, September 1991.
30. Frazier, *Hundredth Monkey*, p.169.
31. Randi, "Help Stamp Out Absurd Beliefs."
32. Martin Ebon, *Psychological Warfare: Threat or Illusion* (New York: McGraw-Hill, 1985), p.232.
33. *The Skeptical Inquirer*, Summer 1991, p.33.
34. "Psychic vs. Skeptic."
35. Hines, *Pseudoscience and the Paranormal*, p.64.
36. Kendrick Frazier, ed. *Science Confronts the Paranormal* (Buffalo, N.Y.: Prometheus, 1986), p.213.
37. Kurtz, *A Skeptic's Handbook of Parapsychology*, p. 141.

38. Hines, *Pseudoscience and the Paranormal,* pp. 27, 28.

39. "Psychic vs. Skeptic."

40. Frazier, *Science Confronts the Paranormal,* p.36.

41. Ibid., p.5.

42. Randi, "Help Stamp Out Absurd Beliefs."

43. Arvey, *ESP: Opposing Viewpoints,* p.40.

44. *Into the Unknown.*

45. Alcock, *Science and Supernature,* p.5.

46. Arvey, *ESP: Opposing Viewpoints,* p.91.

47. Holden, "Irrationality."

48. Ebon, *Psychological Warfare,* pp.226–234.

49. Frazier, *Science Confronts the Paranormal,* p.28.

CHAPTER FOUR:
Battleground: The Human Body

1. Peter Bartrip, "Quacks and Cash," *History Today,* September 1990.

2. "The Top Ten Health Frauds," *Consumer Review,* February 1990.

3. Ira Wolfam, "Special Report: Health Fraud 1990," *New Choices,* December 1990.

4. Ibid.

5. "The Top Ten Health Frauds."

6. Ibid.

7. Wolfam, "Health Fraud 1990."

8. Huber, *Galileo's Revenge,* p.4.

9. Robert L. Taylor, "Beware of Health Hype," *Reader's Digest,* March 1991.

10. Barrie R. Cassileth, "Questionable and Unproven Cancer Therapies," *Consumer Research Magazine,* September 1991.

11. Ronni Sandroff, "Unconventional Cancer Cures: Hope or Hoax?" *New Choices,* October 1991.

12. Cassileth, "Questionable and Unproven Cancer Therapies."

13. Taylor, "Beware of Health Hype."
14. Cassileth, "Questionable and Unproven Cancer Therapies."
15. "The Top Ten Health Frauds," *Consumer Research.*
16. Sandroff, "Uncommon Cancer Cures."
17. Carol Krucoff, "Unlocking the Secrets of Taxol," *Saturday Evening Post,* September 1991.
18. Gene Bylinsky, "The Race for a Rare Cancer Drug," *Fortune,* July 13, 1992.
19. "Vitamin C Gets a Little Respect," *Science,* October 18, 1991.
20. Anastasia Toufexis, "The New Scoop on Vitamins," *Time,* April 6, 1992.
21. "Vitamin C Gets Respect."
22. Toufexis, "The New Scoop on Vitamins."
23. Ibid.
24. "Vitamin C Gets Respect."
25. "Wonder Cures from the Fringe," *U.S. News & World Report,* September 23, 1991.
26. "Why New Age Medicine Is Catching On," *Time,* November 4, 1991.
27. "Wonder Cures from the Fringe."
28. Ibid.
29. Ibid.
30. Ibid.
31. Ibid.
32. Ibid.
33. Wolfam, "Health Fraud 1990."
34. *Into the Unknown.*
35. "Those Biorhythm and Blues," *Time,* February 27, 1978.
36. Ed Nelson, "New Facts in Biorhythms," *Science Digest,* May 1976.
37. "Those Biorhythm and Blues."
38. Ibid.
39. Nelson, "New Facts in Biorhythms."

40. "This Calculator Tells You How You'll Feel on Thursday," *Popular Mechanics*, October 1976.
41. "Those Biorhythm and Blues."
42. "This Calculator . . ."
43. Nelson, "New Facts."
44. "Those Biorhythm and Blues."
45. Ibid.
46. Nelson, "New Facts."
47. "Those Biorhythm and Blues."
48. Susan Perry and Jim Dawson, "What's Your Best Time of the Day?," *Reader's Digest*, November 1988.
49. "Beat the Body Clock," *Mademoiselle*, April 1991.
50. Perry and Dawson, "What's Your Best Time of the Day?"
51. Steve Nadis, "Rhythm & Blues," *Technology Review*, January 1991.
52. Ibid.
53. "Beat the Body Clock."

CHAPTER FIVE:
Battleground: Planet Earth

1. Huber, *Galileo's Revenge*, pp.25–31.
2. *Into the Unknown*, p.66.
3. Ibid.
4. *The Skeptical Inquirer*, Winter 1992, pp.179–180.
5. Frazier, *Hundredth Monkey*, p.355.
6. *The Skeptical Inquirer*, Winter 1992, p.180.
7. Frazier, *Hundredth Monkey*, p.355.
8. Ibid.
9. "The Lesson of Dr. Browning," *Science*, August 9, 1991.
10. *The Skeptical Inquirer*, Spring 1991.
11. "The Lesson of Dr. Browning."
12. Holden, "Irrationality."
13. J. Edward Evans, *Freedom of Religion* (Minneapolis: Lerner, 1988), p.57.

14. Dana Wechsler Linden, "Dreary Days in the Dismal Sciences," *Forbes*, January 21, 1991.

15. Ibid.

16. "Economic Predictions," *The New Yorker*, July 8, 1991.

17. Linden, "Dreary Days in the Dismal Sciences."

18. Ibid.

CHAPTER SIX:
Battleground: The Cosmos

1. "Chandrasekhar vs. Eddington—An Unanticipated Confrontation," *Physics Today*, October 1982; *Current Biography*, 1986.

2. Frazier, *Science Confronts the Paranormal*, p.219.

3. Lawrence Jerome, *Astrology Disproved* (Buffalo, N.Y.: Prometheus, 1977), p.1.

4. *The Skeptical Inquirer*, Winter 1992, p. 173.

5. "Anti-Science Trends in the Soviet Union," *Scientific American*.

6. Julia Parker, *A History of Astrology* (London: Andre Deutsch, 1983), p.7.

7. *Into the Unknown*, p.13.

8. S.J. Tester, *A History of Western Astrology*. Rochester, N.Y.: Boydell Press, 1987, p.18.

9. Parker, *A History of Astrology*, pp.12–14.

10. Hines, *Pseudoscience and the Paranormal*.

11. Ibid., p.141.

12. *Into the Unknown*.

13. Tester, *A History of Western Astrology*, p. 2; Frazier, *Hundredth Monkey*, p.291.

14. *The Skeptical Inquirer*, Summer 1991, pp.378, 379.

15. *Into the Unknown*.

16. *The Skeptical Inquirer*, Summer 1991, p.379.

17. *World Book Encyclopedia*.

18. Hines, *Pseudoscience and the Paranormal*, p.153.

19. Frazier, *Science Confronts the Paranormal*, p.220.

20. Frazier, *Hundredth Monkey*, p.291.

21. Ibid., p.299.
22. Ibid., p.322.
23. *The Skeptical Inquirer*, Winter 1991, p.150.
24. Frazier, *Science Confronts the Paranormal*, p.219.
25. Parker, *A History of Western Astrology*, p.184.
26. Hines, *Pseudoscience and the Paranormal*, p.128.
27. *The Skeptical Inquirer*, Fall, 1991, pp.69–79.
28. Hines, *Pseudoscience and the Paranormal*, p.167.
29. "High-Rise UFO Abductions," *Omni*, April 1992.
30. Michael Arvey, *UFOs: Opposing Viewpoints* (San Diego: Greenhaven, 1989).
31. Ibid.
32. Dava Sobel, "Is Anybody Out There?," *Life*, September 1992.
33. Hines, *Pseudoscience and the Paranormal*, p.173.
34. *The Skeptical Inquirer*, Fall 1991.
35. Hines, *Pseudoscience and the Paranormal*, p.54.
36. *The Skeptical Inquirer*, Summer, 1991, pp.69–79.
37. Sobel, "Is Anybody Out There?"
38. *The Skeptical Inquirer*, Spring 1991, p.255.
39. Sobel, "Is Anybody Out There?"
40. *The Skeptical Inquirer*, Spring 1991, p.255.

CHAPTER SEVEN:
A Pseudoscience Quiz

1. Hines, *Pseudoscience and the Paranormal*.
2. *The Skeptical Inquirer*, Fall 1991, p.162.

EPILOGUE:
The Importance of Recognizing Pseudoscience

1. Huber, *Galileo's Revenge*, p.25.
2. Hines, *Pseudoscience and the Paranormal*, p.19.
3. *Into the Unknown*.
4. Hines, *Pseudoscience and the Paranormal*, p.19.

FOR FURTHER READING

Arvey, Michael. *ESP: Opposing Viewpoints*. San Diego: Greenhaven, 1989.

Arvey, Michael. *UFOs: Opposing Viewpoints*. San Diego: Greenhaven, 1989.

Alcock, James. *Science and Supernature*. Buffalo, N.Y.: Prometheus, 1990.

Frazier, Kendrick, ed. *The Hundredth Monkey & Other Paradigms of the Paranormal*. Buffalo, N.Y.: Prometheus, 1991.

Frazier, Kendrick, ed. *Science Confronts the Paranormal*. Buffalo, N.Y.: Prometheus, 1986.

Hines, Terence. *Pseudoscience and the Paranormal*. Buffalo, N.Y.: Prometheus, 1988.

Huber, Peter W. *Galileo's Revenge: Junk Science in the Courtroom*. New York: Basic Books, 1991.

Randi, James. *The Mask of Nostradamus*. New York: Charles Scribner's Sons, 1990.

Reader's Digest Association. *Into the Unknown*. Pleasantville, N.Y.: Reader's Digest Association, 1981.

INDEX

Page numbers in *italics* refer to photographs.